CIRCUIT CONSULTANT'S CASEBOOK

CIRCUIT CONSULTANT'S CASEBOOK

T. K. Hemingway BSc (Hon)
Stevenage Head of Electronic Technology,
British Aircraft Corporation, Stevenage

BUSINESS BOOKS LIMITED

London

First published 1970

SBN 220.79857.5

This book has been set in 10 on 12 pt Times Roman and printed in England by Hazell Watson & Viney Limited, Aylesbury, Bucks, for the publishers, Business Books Limited (registered office: 180 Fleet Street, London EC4) publishing office: Mercury House, Waterloo Road, London SE1

MADE AND PRINTED IN GREAT BRITAIN

contents

list of figures

introduction

MANY engineers trained in circuit design find that although they feel they have grasped the principles of design, their attempts to put these into practice are disappointing. This book has been written in an attempt to remedy this situation by the use of two approaches. The first part is devoted to the overcoming of typical design hazards; the second deals with techniques which enable unusual requirements to be met by simple means.

It is assumed that the reader is familiar with the fundamentals of circuit design as described, for example, in the companion work *Electronic Designer's Handbook* (Business Books, Second Edition 1970).

The treatment is mainly non-mathematical and the emphasis throughout is on the design of practical circuits: most of the illustrated examples include component values suitable for a typical application.

Inevitably in a book of this type, designed to be useful to a large number of engineers at all levels, some of the material must already be well known to the experienced designer. However, it is expected and hoped that at least some of the techniques described are new to most readers since they represent not merely the author's first-hand knowledge but also the findings of dozens of his colleagues during the last sixteen years.

I wish to add my grateful thanks to those engineers who, knowingly or not, have contributed to this book; to the management of the British Aircraft Corporation, Stevenage, for permission to publish; finally, to Valerie Smith who painstakingly converted my scribble into type.

TKH

PART 1

avoiding design faults

CHAPTER 1

wrong d.c. conditions

FAILURE of a circuit to operate correctly is often caused by phenomena that are difficult to predict by a normal process of calculation. These include certain thermal effects, unintended feedback loops, parasitic oscillations and other happenings of which the experienced designer has learnt to be wary. In this chapter only readily calculable and, therefore, comparatively elementary circuits will be dealt with, emphasis being placed on the avoidance of the kind of error which experience has shown to be most common, namely d.c. conditions which are not as intended.

This situation usually arises from the carelessness brought about by the urge to complete the final stages of design and begin practical tests. This can only be put right by a self-discipline that compels the designer to check and re-check the rather tedious calculations involved. When the design is believed to be complete, it is good practice to calculate the approximate d.c. level, or excursion limits if appropriate, of every terminal of each device in the circuit: this often uncovers hidden errors.

Although the reader is assumed to be familiar with normal transistor circuit design procedure, there are some areas of uncertainty that can lead to errors of the type under discussion. For instance, in designing suitable resistor values for the audio amplifier circuit of Fig. 1.1, one might choose $V_p = 12$ V, $V_c = 8$ V, $V_b = 4$ V, $I_c = 1$ mA, as reasonable working conditions. This immediately fixes $R_3 = 3{\cdot}9$ kΩ, $R_4 = 3{\cdot}3$ kΩ using standard resistor values. Now, in choosing R_1 and R_2, the principle is well known: the ratio must fix V_b to be 4 V, and the actual values must be low enough to give negligible change of V_b when base current flows.

To work this out correctly, the allowable tolerance in V_c for correct circuit operation must first be decided. This depends on the

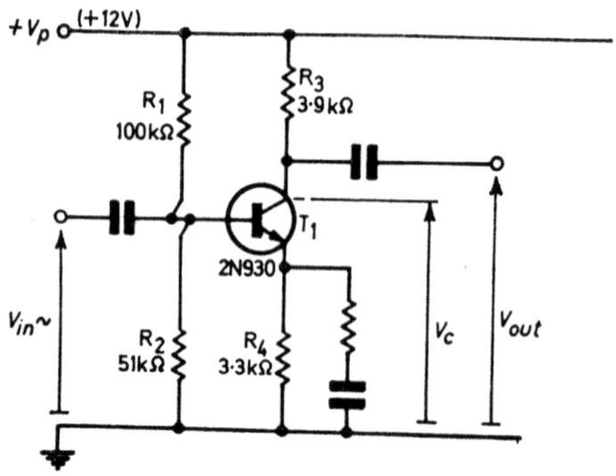

Fig. 1.1 Simple audio amplifier

signal level to be handled, but could in a particular case be 0·5 V. To cause such a change in V_c requires a change in I_c of $0{\cdot}5/R_3 = 0{\cdot}13$ mA, and this in turn requires a change in V_e and V_b of $0{\cdot}13 \times R_4 = 0{\cdot}4$ V. R_1 and R_2 must therefore have a paralleled value such that $I_b(R_1 \| R_2) < 0{\cdot}4$ V, where I_b is the maximum base current possible under the conditions of use of the transistor. In this example β_{min} could be 100, giving $R_1 \| R_2 \leqslant (0{\cdot}4/10)$ MΩ, i.e. $R_1 \| R_2 < 40$ kΩ.

Knowing that $V_b = 4 = V_p R_2/(R_1 + R_2)$ gives the limit values $R_1 = 120$ kΩ, $R_2 = 60$ kΩ. Therefore $R_1 = 100$ kΩ, $R_2 = 51$ kΩ would be safe values.

the base-chain-current fallacy

The above procedure is elementary and rather tedious, but cannot be by-passed. A special warning is in order concerning the often quoted rule that the 'current down the base chain should be many times the base current', a factor of 10 or 20 usually being recommended.

Although rules of thumb of this type are admirable in avoiding unnecessary tedium, there are some, like the one quoted, that are highly dangerous in being based on unsound principle.

Applying the base-chain rule, using a factor of 10, to the present case, the value of $R_1 + R_2$ comes to $V_p/(10I_b) = 120$ kΩ giving $R_1 = 80$ kΩ, $R_2 = 40$ kΩ. $R_1 \| R_2$ is therefore 26·7 kΩ, safer than the 40 kΩ required so that the rule appears to be adequate.

Even this is a slightly questionable result, as the designer may well assume that his factor of 10 on I_b implies a 10 per cent resulting influence of I_b in fixing V_c. If he therefore wishes to allow a full amount of 0·4 V variation in V_c in order to gain the highest value of $R_1 \| R_2$ (because of loading considerations) he would not dare to go higher than the above values; the correct calculation shows they could each be 50 per cent higher.

A much worse danger inherent in this base-chain-current criterion is illustrated in Fig. 1.2. Here, the base supply is derived from ± 50 V lines and this criterion would give $R_1 + R_2 \leqslant (100/0{\cdot}1)$ kΩ or $R_1 + R_2 \leqslant 1$ MΩ, giving the limit values $R_1 = 460$ kΩ, $R_2 = 540$ kΩ. The use of such values would be catastrophic, because $R_1 \| R_2 = 250$ kΩ. The presence of I_b would cause V_b to fall by 2·5 V, i.e. almost cut off. In fact the circuit would still operate but at considerably less than the intended collector current.

The reader might well ask why anyone would consider using such an arrangement as Fig. 1.2 where, apart from the effect under discussion, a small percentage change in V_{p1} or V_n would also have a large effect on V_c.

The answer is that such arrangements are not always presented in such an obvious way, and a similar effect occurs in the circuit of Fig. 1.3. In this case (an actual example submitted to the writer) V_p and V_n were the only rails available. The transistor used could not withstand more than 25 V, and V_c was held at a safe value (even when input signals caused cut-off) by potential divider R_3,R_6. There

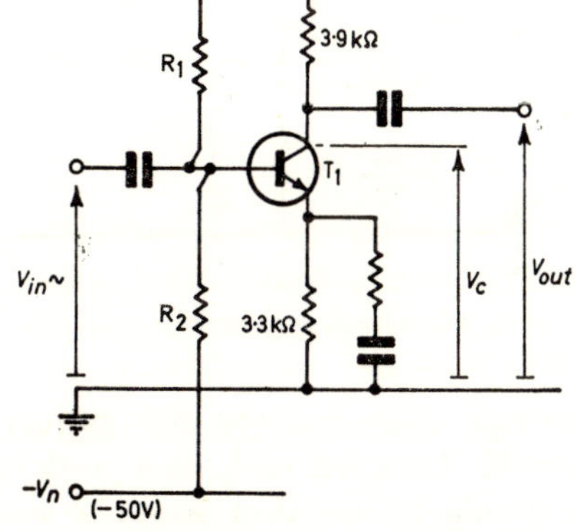

Fig. 1.2 Base supply from high-voltage rails

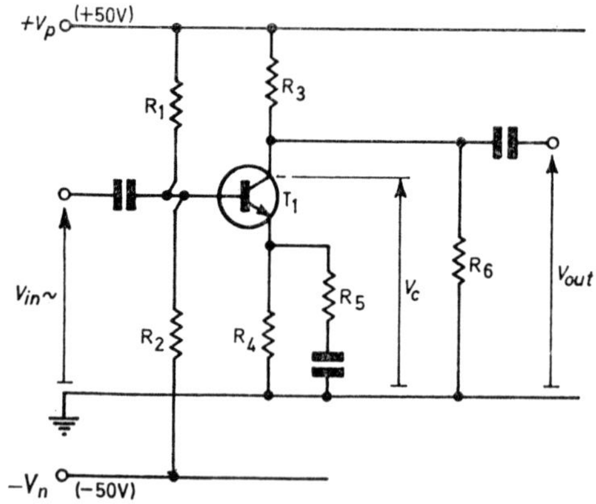

Fig. 1.3 Practical example similar to Fig. 1.2

appeared to be no point in dealing similarly with the base circuit and the use of the negative line instead of earth seemed to allow a higher value for R_2 and therefore less loading of the input signal (!).

The critical point at issue is the resistance value seen from the base terminal, not the (totally irrelevant) current flowing through the chain. The fallacy is easily seen through by examining Fig. 1.4,

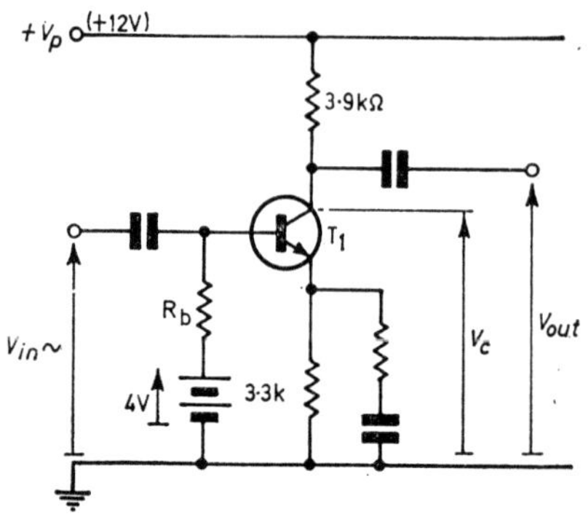

Fig. 1.4 Effect of separate bias supply

where there is no base chain and the only current flowing in the base supply is I_b itself. The effect of I_b on V_c is dependent upon R_b which, as above should be less than 40 kΩ to meet the imposed restriction on V_c. In this circuit, which is a diametrically opposite example to Fig. 1.2, the base current 'factor' is only unity, but the circuit is quite acceptable.

Examples where the base-chain rule gives misleading results are numerous and the pitfall can always be avoided by working out the resistance seen attached to the base and simply multiplying by the base current. Knowing now the change in base voltage, the result can be clearly seen. Simple analysis shows that the base-chain rule gives, for a fixed apparent safety factor of, say, 10, a variable confidence according to the chain supply voltage and the actual level at which the base is intended to sit.

omission of base resistors

A situation commonly arises where one transistor feeds directly into another, as in Figs 1.5a to 1.5e. In the first four circuits a resistor R_{b1} is shown connected to the base of T_2. In the last circuit (Fig. 1.5e) this resistor has been omitted – a very common practice but, as will be seen, one of doubtful validity on no less than three counts. The following comments apply with slight modification to all five circuits but are written specifically for Fig. 1.5e.

In this arrangement the emitter current of T_1 is equal to the base current of T_2 which is itself equal (see Chapter 2, EDH) to (I_{e2}/β_2 − I_{CBO2}). Now if I_{CBO2} should ever exceed I_{e2}/β_2 this current is in the wrong direction for T_1 conduction. T_1 emitter and T_2 base potentials rise until equilibrium is reached when the reverse emitter leakage current of T_1 balances $I_{CBO2} - I_{e2}/\beta_2$. In this state T_2 is conducting but T_1 is cut off; V_{out} is independent of V_{in} and the circuit fails to operate.

When germanium transistors were in common use, most designers rapidly became aware of the problem because of the high values of I_{CBO} involved. With the advent of early silicon devices the danger virtually disappeared because β was low and I_{CBO} negligible compared with I_e/β at normal operating levels. The situation today is, however, reversed again because many modern planar silicon transistors have extremely high β even at very low operating currents. It is quite normal to use for T_2 the well known BC108B at $I_e = 10\ \mu A$, giving a typical I_e/β of about 0·04 μA. If the operating temperature is 90°C, then I_{CBO} can exceed this value and the circuit fails.

When T_2 is a power transistor the danger is considerably greater

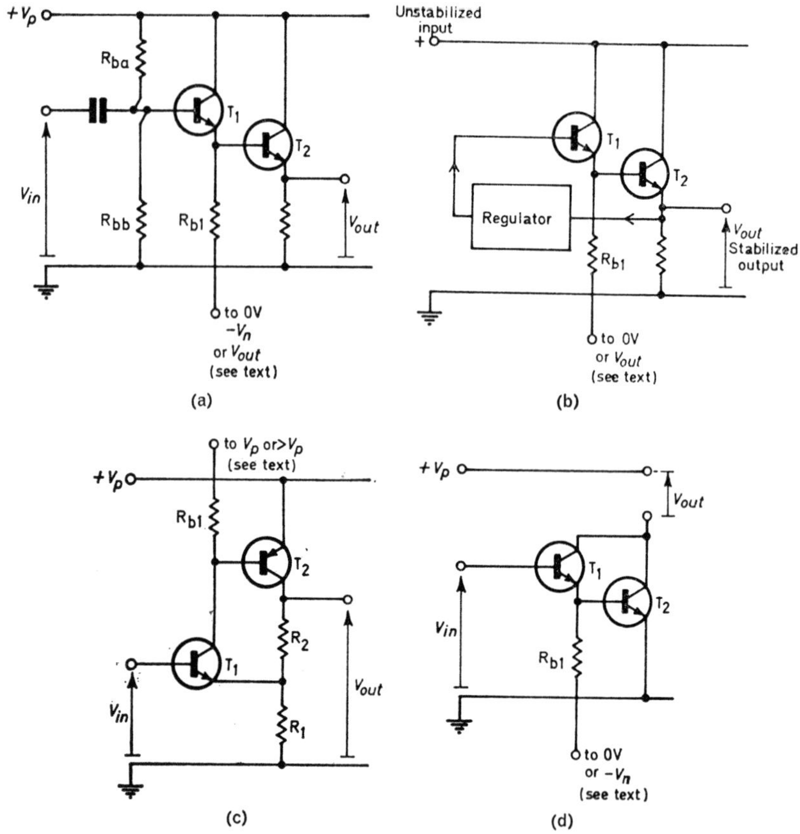

Fig. 1.5*a–d* Use of base resistors: (*a*) cascaded emitter followers; (*b*) stabilizer output drive; (*c*) complementary feedback amplifier; (*d*) buffer and amplifier

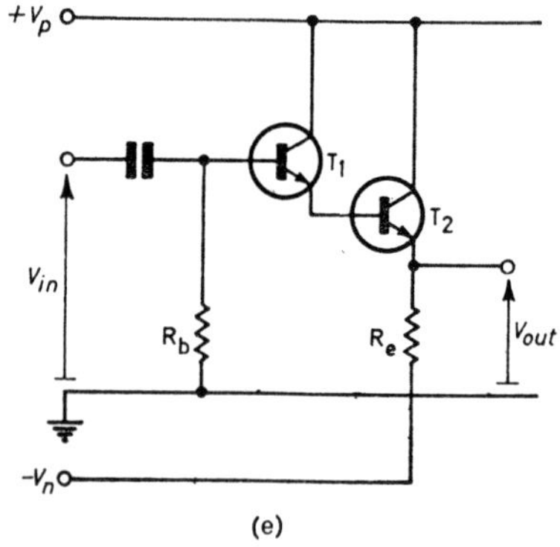

Fig. 1.5*e* Omission of T_2 base resistor

especially in power supply output stages. The 2N3771, a typical silicon power transistor, can have an I_{CBO} of 2 mA even at room temperature; on examining the β characteristic it is clear that circuit failure could occur with any output current *less* than 160 mA. Although one might question the likelihood of using such a large power device for this rather small current it must be remembered that a power supply might be required to supply any current from a few milliamperes to several amperes on different occasions. The greatest danger of all occurs when a heavy load is applied and then removed. When the load is on, the power transistor becomes hot, its I_{CBO} increasing by a large factor. The circuit operates correctly because the forward component of base current I_{e2}/β_2 is also large.

On removing the load, I_{e2}/β_2 falls but I_{CBO} is still at its high-temperature value for some seconds. The base current therefore attempts to reverse and the base potential rises. Depending upon the collector supply voltage, this can cause T_2 dissipation to rise or fall; if it rises thermal runaway can result because the extra dissipation causes a further rise of I_{CBO}. Even if the dissipation falls as T_2 approaches bottoming, the output voltage is excessive and may cause damage to the load.

This frightening series of events is to be taken very seriously since this phenomenon in either its fully destructive form or in a milder version (loss of stabilization) is *easily the most common power supply stabilizer fault.*

Returning to low power signal circuits, another snag in omitting R_{b1} (for example, in Fig. 1.5c) is that T_2 base current may be too small for correct operation of T_1. It is true that in many cases the whole idea is to reduce T_1 emitter current to as low a value as possible in order to make T_1 base current small. This is often a valid reason in d.c. circuits but, as explained earlier, can be dangerous. In a.c. circuits, however, the intention would normally be to obtain high input impedance to T_1 base when applying a small a.c. signal. It is then inappropriate to reduce the current too far, since this causes a drop in the β of T_1.

Yet another situation arises if high-frequency performance is important for either sine or pulse waveforms. When T_1 is being turned on harder by a fast input, T_2 base and its associated capacitance is driven by the available current from T_1 and will follow at an appropriate speed. When T_1 is being turned off rapidly, however, the only drive tending to make T_2 reduce its current rapidly is the path through R_{b1}. Its omission can therefore be expected to cause a poor response time in the direction when T_2 is turning off; a practical check confirms this suspicion.

The consequences of omitting such base resistors are, then, as follows: a danger of complete loss of correct d.c. conditions; too low an operating current for the drive transistor, giving low β; poor and non-linear high-frequency response. They may on occasion be omitted but the right approach is to put them in and regard their proposed removal as case for detailed investigation.

values for base resistors in figs 1.5a to 1.5e

As indicated in the diagrams the base resistor R_{b1} may be returned to the emitter of T_2 or to a steady direct potential. Its value is chosen so that if T_1 is driven to cut-off then T_2 is also cut off, so that in Fig. 1.5c with R_{b1} returned to T_2 emitter, $R_{b1}(I_{CBO}$ for T_1 + I_{CBO} for $T_2) <$ voltage threshold for conduction of T_2 (about 0·5 V). If instead R_{b1} is returned to a higher voltage than V_p, voltage V'_p, then the condition is

$$R_{b1}(I_{CBO1} + I_{CBO2}) < V'_p - V_p + 0{\cdot}5$$

The advantage gained is that for given values of I_{CBO1} and I_{CBO2}, the safe limit for R_{b1} is higher thus giving less signal attenuation. The use of an extra supply line in this way is hardly ever justified on economic grounds, since the improvement in loop gain due to a

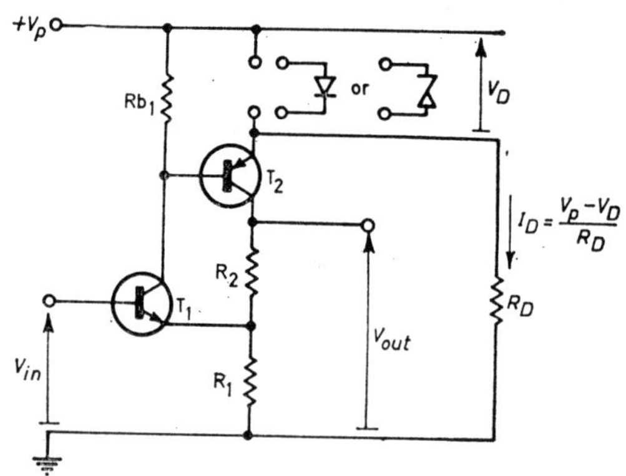

Fig. 1.5f Creating an extra supply to T_2 emitter

larger R_{b1} is only marginal; in a large system, however, the line may already exist. In using germanium transistors with high I_{CBO} and a turn-on threshold of less than 100 mV, R_{b1} does represent a heavy load on T_1 as its value is much lower. In this situation an extra supply may be worth creating even if it is only 0·5 V higher than T_2 emitter potential. A convenient method to achieve this is to return R_{b1} to V_p and to insert a silicon diode in T_2 emitter lead, giving T_2 an apparent threshold of about 0·6 V. Alternatively, a Zener diode of several volts breakdown may be used for still larger R_{b1}; in either case it is advisable to bleed current into the diode as shown in Fig. 1.5f, thus keeping T_2 emitter circuit impedance low to maintain circuit gain.

d.c. conditions in high input impedance circuits

The circuit shown in Fig. 1.6 is widely used and gives a well defined gain $(R_4 + R_3)/R_3$ with high input impedance (see Chapter 13). The input direct current required by T_1 base with the values specified in Fig. 1.6 is of the order of 1 μA and a typical specification might call for an input impedance of 1 MΩ and a d.c. drift referred to the input of a few millivolts.

Some of this drift is given directly by the inequality of the V_{be} of T_1 and T_2 and by the fact that the loop gain of the circuit is not infinite. The point mainly to be considered here, however, is the effect of the 1 μA input current, which gives an input offset voltage equal to $R_S \times 1\ \mu$A. Therefore the drift depends directly on the

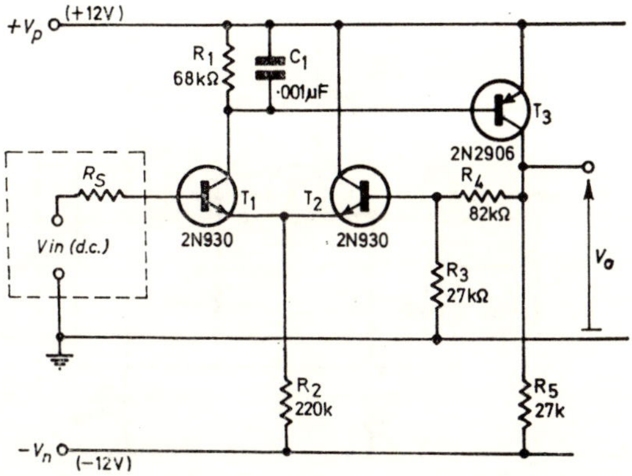

Fig. 1.6 D.C. feedback amplifier with high input impedance

source resistance R_S and not at all on the input impedance of the amplifier. The designer must, then, be told what source resistance is used as he cannot guess whether this is in the region of 200 kΩ (allowing some 20 per cent loss at the input) or perhaps 1 kΩ (allowing only 0·1 per cent loss). The tendency is to think of the 1 MΩ as the source resistance and assume a drift of about 1 V.

A similar, but slightly less obvious, error can arise when the specification for such an amplifier also has a certain drift requirement even with the input disconnected. Under these conditions the circuit given would certainly settle at very different levels from its intended working conditions. The only hope would be to add a resistor of, say, 1·2 MΩ from T_1 base to ground, giving a typical drift of 1·2 V with the source removed and still having an input impedance of several hundred kilohms. If this drift is outside the specification (which is very likely!) the circuit would have to be redesigned using lower currents, or a matched pair of field effect transistors for T_1 and T_2.

This is all straightforward; the pitfall appears if the designer conceives the idea of bootstrapping, a technique which is most useful in raising the input impedance of such an amplifier whilst retaining low value base resistors (Fig. 1.7). Here R_1 may have quite a low resistance, e.g. 22 kΩ ,but because its lower terminal is forced by the unity gain positive feedback to move by just the same amount as its upper terminal, its apparent value is infinitely high, i.e. it takes no current when the signal level changes. In practice, $R_3 : R_2$ and $R_4 : R_5$ do not match precisely, and V_{R2} and V_{in} are not exactly

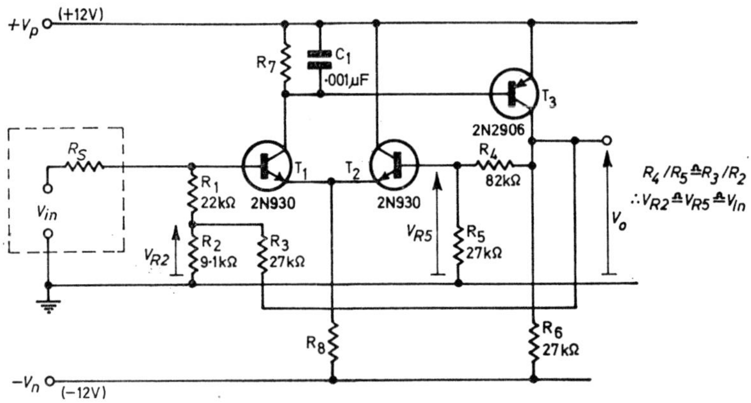

Fig. 1.7 Bootstrapped feedback amplifier

equal since the loop gain is not infinite. Even so, an input impedance of 1 MΩ is not difficult to achieve.

This technique is often used in one of several a.c. coupled arrangements (see Chapter 15 of EDH) and still works successfully when directly coupled.

The fallacy arises in believing that drift, with no source connected, is improved by obtaining the high input impedance in this way rather than using a simple input resistor to ground of the required high value.

This is not true, because R_1 appears to have a high value from the point of view of the input base current and this high value determines the resulting drift. Bootstrapping in this directly coupled form is hardly ever useful; it does not improve drift since the low value of R_1 itself is not relevant, and the apparent value of R_1 which determines drift and input impedance is badly defined.

non-starting d.c. circuits

There are some well known circuit configurations which, although often taught as 'standard' circuits, and which appear to be satisfactory when analysed, fail to work at the intended d.c. levels.

The next chapter deals with circuits that fail permanently due to damage caused by various transients, although safely designed for normal working conditions. The present section is concerned with non-destructive circuit failure which occurs when the intended operating conditions, correct in themselves, are not necessarily reached.

conventional stabilizer

The circuit of Fig. 1.8 is a simple voltage stabilizer in which the Zener diode ZD_1 operates from the current in T_1 plus the current from R_4. At switch on, if the output is assumed to stay at zero, no current is supplied to ZD_1 from R_4, or from T_1. However, R_1 passes a current $(V_{in} - V_{beT_2})/R_1$ into T_2 base (since T_1 is cut off) and this causes T_2 to turn on hard; V_{out} rises and ZD_1 is brought into conduction. This process continues until T_1 is turned on when voltage at the R_2,R_3 junction exceeds V_{ZD_1} by V_{beT_1}. The output becomes stabilized at a voltage of $(V_{ZD_1} + V_{beT_1})(R_2 + R_3)/R_3$

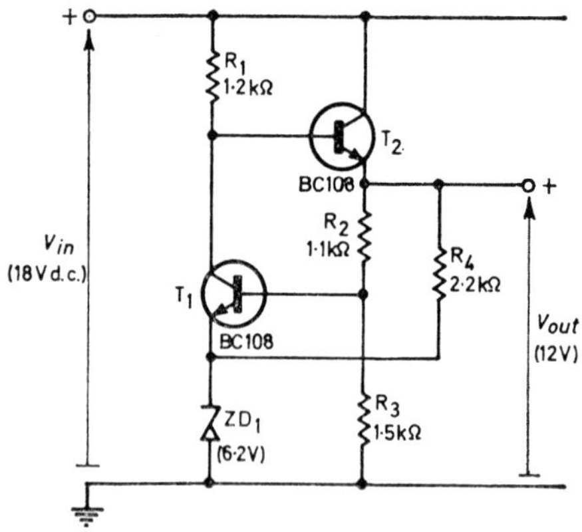

Fig. 1.8 Simple stabilizer

at this point. This stabilizing sequence still applies if R_4 is omitted but certain modifications cause failure-to-start as shown in the next section.

improved stabilizer with starting difficulties

One of the snags of this simple circuit is that supply line variations are coupled through R_1 directly to T_2 base, thus influencing the output. These variations are naturally reduced by the gain of the loop, but this factor is not very high in this simple circuit. The circuit in Fig. 1.9 attempts to remove this effect by using a *pnp* output transistor so that supply line variations affect its base and emitter equally, causing no change in its collector current and therefore no output variations. At the same time, to keep the phasing of the loop correct, i.e. *negative* feedback, an extra transistor is introduced.

It is easy to check that all values are easily designable and that the circuit should function. At switch-on, however, if V_{out} is zero, T_1, T_2 and ZD_1 are cut off. The circuit has been carefully (!) designed to ensure that the voltage on R_1 does not affect the conduction of T_3 since its emitter is also at $+V_p$. The circuit will therefore not 'start' unless some transient causes V_{out} to rise momentarily and turn on ZD_1, after which normal operation will take place.

Any circuit involving direct couplings should be examined in this way, on the assumption that it may not start. If the circuit really is

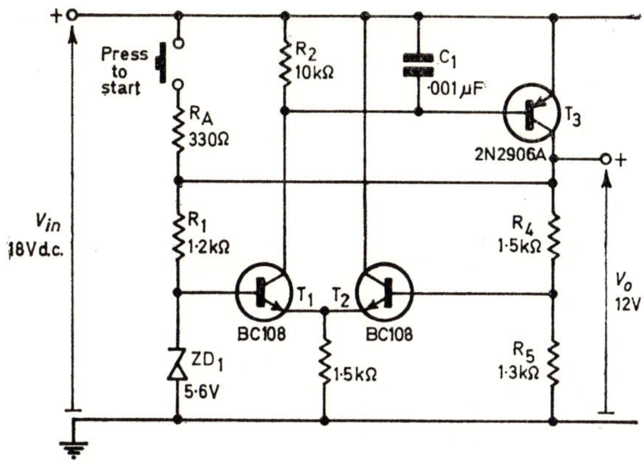

Fig. 1.9 Improved stabilizer for supply ripple rejection

self-starting this *reductio ad absurdum* procedure will prove it by a contradiction, as in the first stabilizer example (in which V_{out} could not after all remain at zero, as originally assumed).

It is easy to find a cure for the fault on the circuit of Fig. 1.9, provided the danger is known. The nature of the cure will depend on the application of the stabilizer. In some cases such as a bench power supply, a 'start button' can be added, as in Fig. 1.9, which starts the Zener on being pressed. Note that such a supply is proof against a short-circuit load since if V_{out} goes to zero, every transistor cuts off rapidly and the supply remains at zero.

In a supply used in the heart of a large equipment, or in a situation where high reliability is vital, it is usually impossible to resort to a push-button start system and self-starting must be built-in. The temptation which must be resisted is to rely on the impulse produced as the d.c. supply to the unit is switched on. Although use could be made of this by, for example, adding a capacitor from the upper terminal of ZD_1 to the input supply, this is a very doubtful procedure.

One danger is that there may be no guarantee that the rate of rise of supply will be fast; it may differ according to whether actual- or test-supplies are used. The other more serious point is that a momentary short-circuit load, an input transient caused by the operation of an overload trip, or even by a lightning flash, could cause the output to fall to zero for a few microseconds. The circuit would then remain in this state until a suitable triggering transient appeared. To restore normal output it would be necessary to shut down and switch on again a large system with considerable time loss.

The starting arrangement must therefore operate in such a way that the output cannot remain at zero, whether or not a transient occurs.

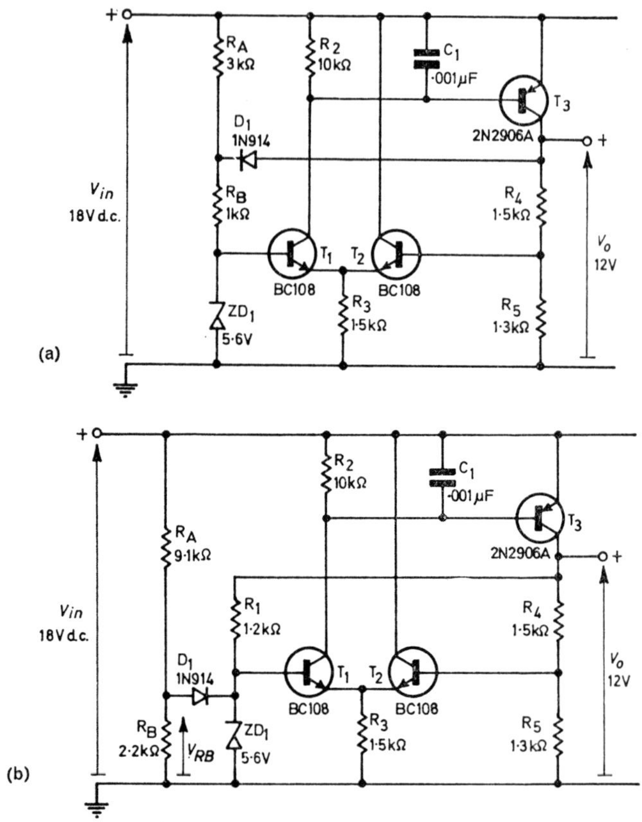

Fig. 1.10 Starting arrangements for circuit of Fig. 1.9

There are many solutions and two are given in Fig. 1.10. The method given in Fig. 1.10a is based on the observation that, had the Zener diode been driven direct from V_{in}, starting would be certain, but variations on V_{in} would of course affect the reference voltage. The principle used is to start the diode as suggested but to arrange that the potential at the junction of R_A and R_B is lower than the normal value of V_{out}. When V_{out} becomes established, diode D_1

then conducts and V_{out} supplies most of the current in R_B. The influence of V_{in} is much reduced because the impedance of the diode, in series with the very low output impedance associated with V_{out}, acts with R_A as a potential divider.

An alternative method (Fig. 1.10b) uses two resistors R_A and R_B to produce voltage V_{R_B} which must be no greater than V_Z, but large enough to turn on T_1, via D_1. At switch-on V_{out} rises, but not to its correct value since its reference voltage ($V_{R_B} - V_{D_1}$) is low. This results in T_1 base voltage rising, because of R_Z, and finally V_{T_1} base reaches V_Z; D_1 is cut off and normal stabilization follows.

The second arrangement is more attractive in some ways since virtually complete isolation from V_{in} is guaranteed when the stabilizing level is reached. In designing for a specific case, however, ensuring suitable values for V_{R_B} in the two circuits for all required values of V_{in} usually leads to a preference for one or other arrangement, depending on the specific values of V_Z and V_{out}.

other non-starters

Many other well known circuits are potential non-starters, either because the d.c. conditions have two or more possible values—only one of these being correct—or because the design relies on a dynamic situation that may never come about. In either case the only good remedy, if a 'start' button is impractical, is to make the circuit correct itself, as pointed out in the last section, rather than rely on a switch-on surge.

By far the most commonly used non-starting circuit is the text-book multivibrator, dealt with in Chapter 9. In system design, a.c. coupled logic circuits and heterodyne loops often fail to operate for basically similar reasons; these are discussed in Chapter 6.

rectification effects

In calculating d.c. conditions the designer should consider the possible influence of signal voltages and currents on the mean operating levels. It is obvious enough that a rectifier circuit used to monitor the level of an alternating signal will produce a d.c. output and may change the bias conditions. This is hardly ever overlooked if the rectifying circuitry is deliberately incorporated for that purpose. There are some circuits, however, that act as rectifiers even though this is not the reason for their existence. This additional property may be overlooked with disastrous effects on circuit performance.

The circuit of Fig. 1.11 illustrates a typical, if simple, error of

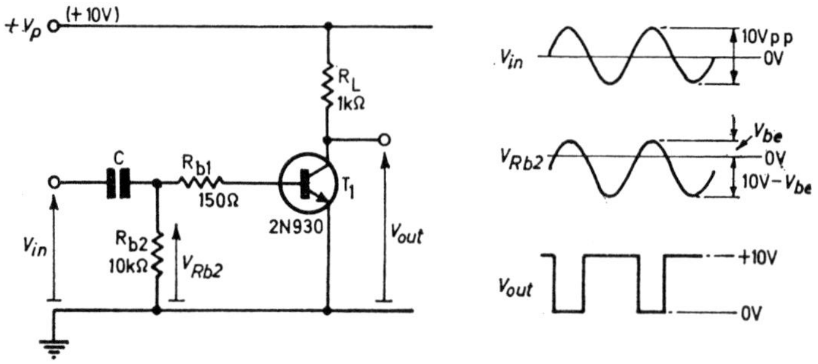

Fig. 1.11 Unsatisfactory squaring circuit

this kind in which one fundamental mistake leads to two unfortunate results. The intention is to generate an approximate 1:1 square wave from an input sine wave of 5 V peak. With so large a signal T_1 would turn on and off more or less symmetrically if biased to 0 V and the signal superimposed through a very large capacitor C. Moreover the reverse drive of about 5 V maximum would be safe for the reverse V_{be} rating and the inclusion of R_{b1} (150 Ω) would limit forward base current to 30 mA.

The performance obtained is found to be reasonably close to original expectation if source resistance is high, e.g. 50 kΩ; but with normal sources of 50, 75, or 600 Ω the output mark/space ratio differs very considerably from unity, the mean output level being close to 10 V. Worse yet, the transistor tends to fail catastrophically when a 50 or 75 Ω source is used.

The mistake made here is the assumption that the signal across R_{b2} is symmetrically disposed about zero because of the a.c. coupling through C and the earthy return of R_{b2}. On first applying the input this condition does hold, but when a positive half cycle occurs, T_1 base emitter diode conducts and causes an extra charging current in C from left to right (in Fig. 1.11) through R_{b1}; on the negative half cycle T_1 cuts off and a charging current passes into C from right to left only through R_{b2}. When the source resistance is high these currents are almost equal but for low source resistance the positive half-cycle current is much the greater since the path through R_{b1} and T_1 base emitter is of much lower resistance than R_{b2}.

Thus the capacitor C is gaining more charge on positive half-cycles than it loses on negative half-cycles. This is not an equilibrium condition and must cause C to acquire a steady potential in such a direction as to nullify the net charge gained or lost in each complete cycle. Therefore, the right-hand terminal of C will become negative so as to reduce the time during which the larger current is flowing. This effect is often known as d.c. restoration.

In the limit when $R_{b2} \to \infty$ and $R_{b1} \to 0$, the mean level at T_1 base will be $(V_{be} - 5)$, so that the positive tips of the input signal marginally reach T_1 base conduction level. In a real circuit, as shown, T_1 will conduct for a short time near the peak of the input signal giving the far-from-unity mark/space output described above. Note that the negative tips reverse bias T_1 base-emitter junction by almost 10 V, explaining the occasional transistor failure.

Various remedies are possible in this particular circuit; to avoid this trap in other circumstances it is advisable to ensure that capacitors cannot accumulate charge during a signal cycle, if it is intended that mean d.c. levels sbould remain as defined by resistors only. Where the very nature of the circuit tends to produce this effect, regardless of designed values, additional components may have to be added to equalise the unlike charging paths.

To improve the squaring circuit of Fig. 1.11, the circuits in Fig. 1.12 are all effective and are based on the above principle. The first (Fig. 1.12a) merely creates through D_1 an identical charging path for C on negative half-cycles. In some circumstances, the reduction in reverse base drive caused by this diode can prevent correct operation, e.g. chopper transistor base drive (see page 168 of EDH). In these cases Fig. 1.12b may be useful in preserving full reverse drive but still maintaining equal charging paths if $R_{b3} = R_{b1}$. Figure 1.12c shows a different approach in that the positive half-cycle current is greatly reduced so that the current taken by R_b dominates. Another method is to make R_{b2} in Fig. 1.12a much less than R_{b1}.

Much more subtle examples than the above occur when designing the input couplings to chopper amplifiers. This subject is discussed in Chapter 4 and calculation is straightforward – the main point is to recognise during the design phase that this problem may exist whenever a capacitor is used.

Finally, note that the size of capacitor does not affect the equili-

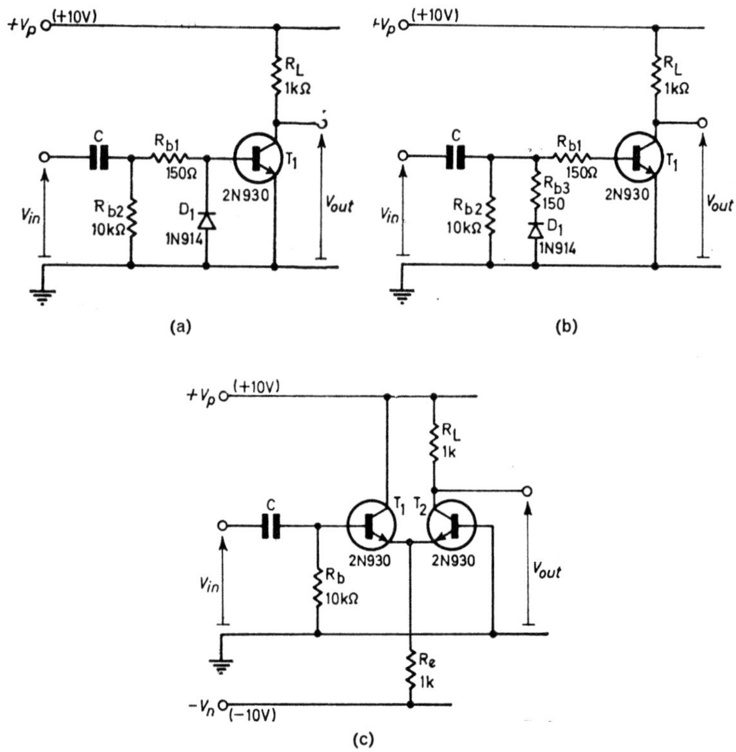

Fig. 1.12 Improved squaring circuits

brium condition provided it is so large that the alternating voltage across its terminals is negligible. Bigger capacitors merely prolong the time taken to reach the equilibrium condition.

parasitic oscillation affecting d.c. levels

Where d.c. conditions are apparently defying the laws of physics, parasitic self-oscillation of transistors is often the only explanation remaining. With modern high f_T transistors used in circuits of quite low frequency, the monitoring equipment may be unable to detect the oscillation, typically in the 500 MHz region.

The magnitude is often a few volts peak and can completely change the operating levels of the transistor in question and others connected to it. Suspicion that this phenomenon is responsible is usually aroused by changes in d.c. level that occur on physically approaching the circuit; and by transistors being apparently cut off (base-emitter being reverse-biased) yet running at a high temperature. This is explained in Fig. 1.13, where it is seen that stray capacity

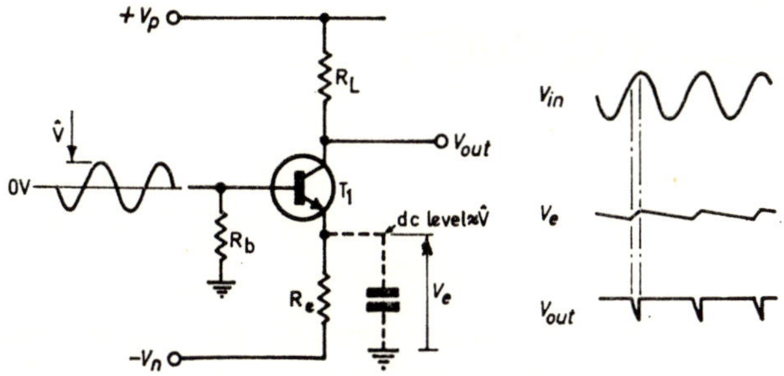

Fig. 1.13 Bias change due to parasitics

on the emitter of an oscillating transistor will cause rectification (see last section) and the current pulses which recharge this capacity may have a high r.m.s. value and give high collector dissipation.

The cure is often difficult but usually a few tens of ohms directly in series with the base lead, and the use of very short transistor leads, will be effective. Anti-parasitic ferrite beads are sometimes (but only occasionally) useful when threaded on one or other transistor lead but it must be remembered that these beads are slightly conductive and may cause unwanted direct current paths.

CHAPTER 2

protective circuitry

WHEN the circuits of well-proven items of electronic equipment are examined, it soon becomes clear that the simple standard forms of circuit usually described in design textbooks are rarely adopted in production designs. Additional components are used, some of them for the obvious purpose of modifying the circuit action in a particular way; others appear to perform no useful function and indeed have no influence at all on the normal mode of circuit operation. In such cases, if the design is well conceived, the extra components are usually present to afford protection to the more vulnerable parts of the circuit. This is often necessary in semiconductor circuits where a transient of only short duration can cause permanent damage.

In finalising a design, consideration must always be given to the possible conditions under which abnormal voltages or currents can occur. This usually requires a careful study of the following effects: switch-on and switch-off conditions; excessive amplitude of input signals; excessive signals which may be transmitted back from an external load; short-circuits on the output.

There may also be dangerous conditions during the normal operation of the circuit, necessitating further modifications from basic text-book circuits.

Care must naturally be taken that the arrangement and value of any added components have the smallest possible effect on the intended mode of operation. Some typical problems are dealt with in the following sections, and the suggestions given should be regarded not as the only, or the ideal solutions, but as a guide to the approach needed.

need for protection components

The simple circuit illustrated in Fig. 2.1 is quite conventional except for the connection of the decoupling capacitor C_2 to the supply line rather than to earth. This would have no appreciable effect on normal performance if the supply had a low impedance over the band of

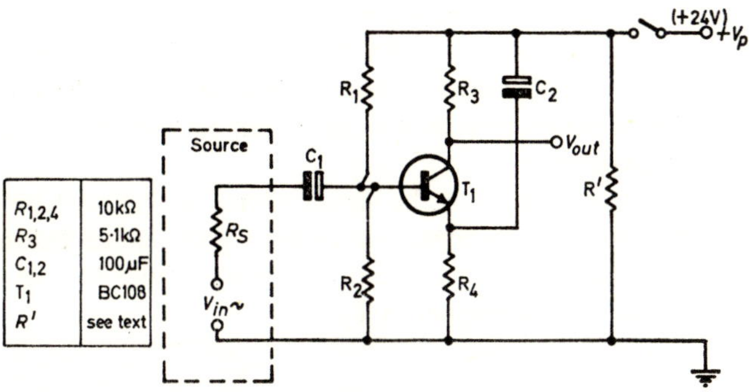

Fig. 2.1 Low-frequency amplifier

frequencies concerned. It would seem then, in view of the ensuing difficulties, that this position of the capacitor has little to recommend it. The designer could, even so, have his reasons; for instance, he may require that T_1 be cut-off for a short time after switching on the supply, for the benefit of a succeeding stage.

switching on

Assume that the signal source v_{in} has a low resistance, e.g. $R_S \approx 100$ Ω, and is connected before switching on V_p. If V_p is now applied suddenly, T_1 emitter immediately rises to $+V_p$(24 V) and T_1 base remains near 0 V, because both C_1 and C_2 have initially zero charge. T_1 emitter-base junction is therefore reverse-biased by some 24 V, causing failure of T_1.

If, on the other hand, the supply is switched on before connecting the signal source, the transistor may survive until the signal source is connected since the base current is limited by R_1R_2. Then, the base-emitter junction is reverse-biased, this time by about 12 V until C_1 charges, again causing transistor failure.

switching off

If normal operating conditions have been achieved without mishap, for example by connecting the source first and raising the supply slowly to +24 V, there is still danger when switching off. A simple open-circuiting of the supply may be harmless if the circuit is considered in isolation. Often, however, there is additional shunt loading R' on the supply which remains attached to the circuit at switch-off. This causes the rapid fall of V_p to zero, tending to pull T_1 emitter to −24 V, whilst its base is held at +12 V by C_1. The resulting base-emitter current flow may well exceed the rating and wreck the transistor.

Previous disconnection of the signal source makes switching off a safe operation; but reconnection of the signal source before the charge on C_1 has leaked away can still cause failure for the same reason—excessive base current in T_1.

basic cause and remedies for switch-on and switch-off damage

Most of the above troubles are caused by the connection of C_2 to the positive supply instead of earth. There is sometimes a good reason for such an arrangement which may become clear only by examining all the associated circuitry, but in almost all applications the more normal connection to earth should be used. This removes the more violent effects, but failure can still occur if the input is removed and subsequently connected after the charge on C_1 has leaked away.

Even if C_2 has to remain in its dangerous position, it is nevertheless still possible to make the circuit safe.

In this example the base-emitter junction is the weak spot as it can become too heavily biased in either direction. Series resistance offers considerable protection in reducing forward current to a safe level.

At switch-off with the source connected, the base, as described earlier, remains momentarily at +12 V and the emitter tries to reach −24 V. The resulting peak current is approximately

$$(36 - V_{be})/R_S$$

This can be limited to a level of 0·1 A by adding 360 Ω in series with the base lead, or by ensuring that the source resistance R_S is of this order.

At switch-on, this protection is less effective because the emitter reaches +24 V and the emitter-base junction starts to conduct heavily in the reverse direction when V_{eb} is about 7 V. C_1 is at this time uncharged, so that the resulting base-emitter current is 17/360 A or about 50 mA, giving a power dissipation in the base-emitter junction of 350 mW. Whether this would cause failure depends on the values of C_1 and C_2 which determine the duration of the effect, and the power rating of the base-emitter junction (not necessarily the same as the total power rating).

Increase of the protective resistor to about 1 kΩ should ensure safety with most transistor types but circuit gain begins to suffer noticeably, since the input impedance at T_1 base is of the same order.

A clearer solution is to add a shunt diode D_1 as well as the series resistor R (see Fig. 2.2). This limits the reverse voltage to a level

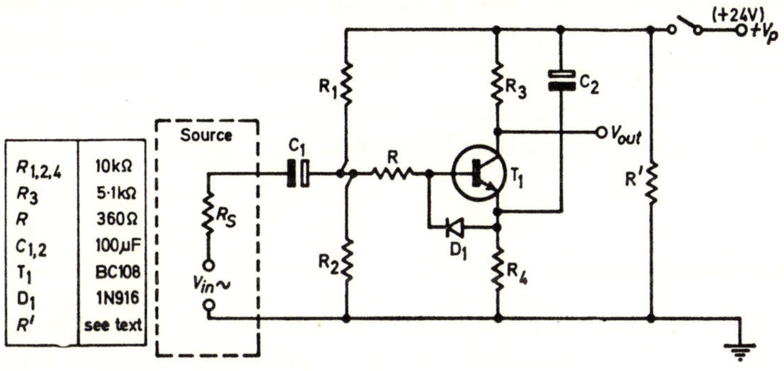

Fig. 2.2 Circuit of Fig. 2.1 protected for switching on/off

which will not allow base-emitter reverse conduction; the diode itself is protected, like the forward base-emitter path, by the resistor.

In this way switching on and off cannot damage the transistor provided the value of R ensures that the absolute base current rating is not exceeded.

An alternative, and apparently equally effective, protection is to insert a diode in series with either the base or emitter of T_1 as shown in Fig. 2.3, so that although normal forward-biased operation is hardly affected, the diode prevents the base-emitter from being subjected to reverse bias while resistor R still protects against excessive forward bias.

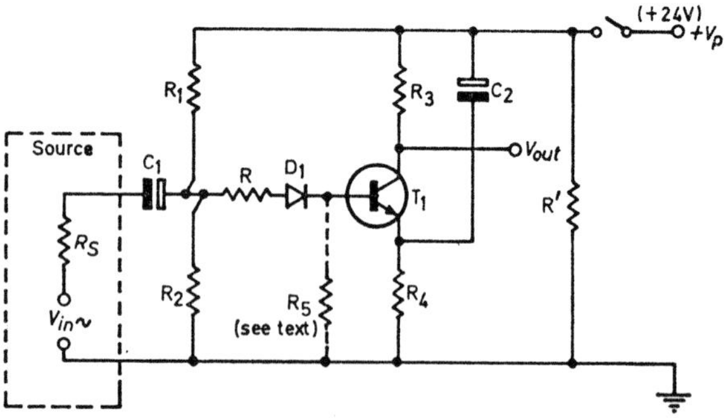

$R_{1,2,4}$	10kΩ
R_3	5·1kΩ
R	360Ω
$C_{1,2}$	100μF
T_1	BC108
D_1	1N916
R'	see text

Fig. 2.3 Alternative protection for circuit of Fig. 2.1 (*but see text*)

One snag is that temperature variations now have more effect on the transistor operating conditions because the potential on T_1 base varies with the forward drop of D_1. Another temperature effect is that if T_1 has high β and also high I_{CBO}, its required base current can be outwards, cutting off D_1 and giving the wrong collector current and voltage. R_5 should therefore be added and of such a value that the current in it exceeds the I_{CBO} of T_1 at the maximum junction temperature of operation (see Chapter 1).

input signal problems

Even now, with the extra diodes and resistors, shown in Fig. 2.2, there is still no protection against a prolonged excessive input signal, e.g. 240 V mains. It is true that the base-emitter junction is now well protected for normal transients but with really large inputs *R* is too small to limit the current effectively. A solution is to add two more diodes D_2D_3 (Fig. 2.4) which clip the base signal, and therefore the transistor currents, to safe levels. It is necessary to split R roughly as shown, because if the diodes are joined directly to R_1,R_2, they would themselves have to deal with unlimited signal currents when clipping. If, on the other hand, they are joined to T_1 base, dangerously large currents can flow in the base-emitter circuit when the clipping level is reached on D_2 and D_3. Current ratings for D_2 and D_3, and the voltage and ripple current rating for C_1, must be sufficient for the maximum expected input signal.

The power supply must be able to accept a reverse load current,

$R_{1,2,4}$	10kΩ
R_3	5·1kΩ
$R/2$	180Ω
$C_{1,2}$	100μF
T_1	BC108
$D_{1,2,3}$	1N916
R''	see text

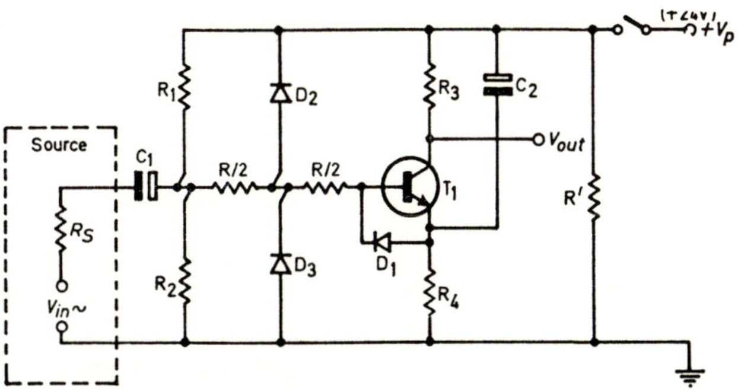

Fig. 2.4 Input protection for circuit of Fig. 2.2

since large positive input half-cycles force current into the supply line. This may require special protection circuitry in the power supply to prevent its output rising in sympathy, causing excessive V_{ce} to T_1 and danger to the supply itself.

Protection against large input can be provided in the same way for the alternative circuit of Fig. 2.3.

dangers from the output load

In certain circumstances the external load on the output of a circuit may itself generate and send back large transient voltages and currents. When the load is under the designer's control this problem hardly exists, but such possibilities must always be considered when the nature of the load is unknown.

In equipment to be used with telephone systems (to give a particularly violent example), a low power audio generator output stage may find itself connected directly to a ringing-tone signal consisting of a 50 V square wave from a low-impedance source.

This could rapidly demolish T_1 in Fig. 2.2 either on account of excessive V_{cb} if driven 50 V positive, or excessive collector current if driven 50 V negative.

remedy for load effects

The problem is similar to the switch-on/off hazard and can be dealt with by adding D_4 and D_5 (see Fig. 2.5) and an extra resistor R″ to protect the new diodes from excessive forward current.

The problems of external load short-circuits are most evident in power amplifier and stabilizer design and are discussed in Chapter 8.

effect of protection circuits on normal performance

The ideal protection device would have no influence on normal circuit operation, and the designer should always design protection circuits with this aim in mind.

Inevitably, performance is changed in some degree and the following indicates the disturbance to be expected from the arrangements considered above.

On examining the 'fully' protected circuit of Fig. 2.5, and beginning at the input, the first effect is the increased input capacity due to

$R_{1,2,4}$	10 kΩ
R_3	5·1kΩ
$R/2$	180Ω
$C_{1,2}$	100μF
R''	180Ω
T_1	BC108
$D_{1,2,3,4,5}$	1N916
R'	see text

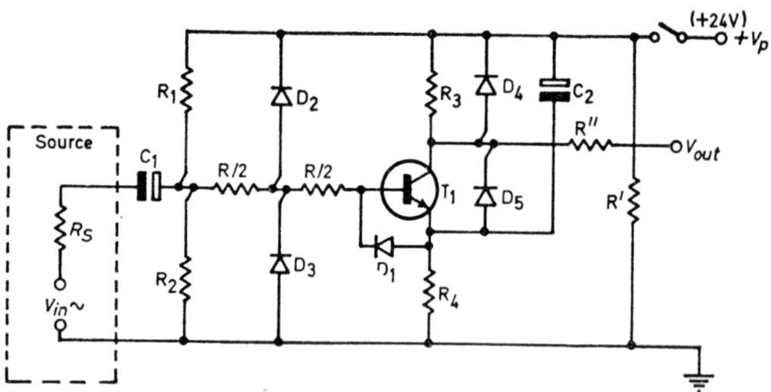

Fig. 2.5 Output protection for circuit of Fig. 2.4

D_2 and D_3. This apparently trivial effect can mean the difference between success and failure in a high-frequency pre-amplifier, such as the *Y* input amplifier of an oscilloscope. In that case T_1 may be a field effect transistor, but the protection principles still apply and an input of several hundred volts is by no means unlikely. At the same time a few extra picofarads can be intolerable from the point of view of instrument specification.

Another point is that when an excessive input is applied, the circuit is protected but the source is not: as clipping occurs the input impedance of the circuit falls to approximately $R/2$ in series with the reactance of C_1, possibly damaging the source. The argument cannot necessarily be used, that such a large input implies in any case a faulty source – it may be, for example, that a wrong switch position was chosen.

The leakage currents of D_2 and D_3 can conceivably upset the bias chain R_1R_2. This is extremely unlikely in a good design and would imply a very high diode leakage, unless T_1 is a field effect device; in this case R_1 and R_2 may be large. This consideration can be important in pre-amplifiers of ultra-high impedance where special low-leakage devices would have to be specified. D_4 has only the very small effect of extra leakage and capacitance between T_1 base and emitter. Once more, this can upset the performance if T_1 is a field effect device.

The output protection diodes D_4 and D_5 add capacitance collector circuit; the effect of diode leakage here will be negligible.

The additional resistors R and R″ both have the effect of lowering the stage gain. R also causes a drop in HF response due to the 'Miller' feedback capacitance C_{cb}; this will be unimportant if $R \ll R_S$ and can in any case be restored by a capacitor in parallel with R, small enough to maintain the required protection. R″ increases the output impedance but this is negligible if $R'' \ll R_3$.

Protection arrangements for power stabilizers can have considerable influence on normal performance and these are discussed in Chapter 8.

multiple supply lines

It is convenient in many circuits, especially those which are part of large systems, to use several positive and negative supply lines.

In such circuits particular care must be taken that the presence or absence of some of the supplies cannot cause damage.

The practice of specifying in which order lines must appear and disappear is not to be recommended; even worse is to specify simultaneous connection.

In the first case, although the supplies of a complete equipment could be designed in the required manner, there are numerous possibilities during development work and in final circuit testing where one or more of the supplies could fail, thus damaging the rest of the circuit. In the second case, there is no real possibility of true simultaneity in switching on two or more supply lines: it is then a matter of luck whether the lag obtained in practice is sufficiently long to cause damage.

The only solution is to add protective measures so that no combination of present and absent supplies is to be feared.

when to use protective components

It is unnecessary to use the full protective measures, such as those described above, in every stage of a circuit chain. These drastic steps are usually of extreme importance at those interfaces between systems where the designer has no control. Typical examples are the input to an oscilloscope and the output of a signal generator or power supply, if these leave the designer as entities. However, a display system in which a generator is coupled permanently to an oscilloscope may require protection only against switch-on and switch-off, since it can be assumed that the oscilloscope and generator can be designed so as not to cause mutual damage.

The designer should therefore look at the danger points where his circuit system begins and ends and whenever external unknown signals or loads can have damaging effects. In addition, switch-on and switch-off surges must always be examined and, of course, the possible dangers of the intended mode of circuit operation.

A quick pointer to possible danger in the last two categories is often provided by noting current paths in which no obvious limit exists to the magnitude of current that can flow. Figure 2.1, for example, has a current path from V_{in}, through C_1, the base-emitter junction of T_1, through C_2, to the supply line, limited only (from the transient point of view) by R_s, which could be virtually zero, if outside the designer's control. Sometimes there is no clear way in which such a current could be initiated. The designer should then be guided by the well-known principle; if a disaster can happen only by the remotest chance, that chance will turn up at the worst possible moment.

Many examples occur in the following chapters where protection is advisable and the problems are discussed as an essential part of the design. Use of the following check-list will avoid the issue of designs that are inadequate in this respect.

check list for protection

1 Look for unlimited current paths.
2 Calculate reverse voltages and instantaneous currents and power that will occur immediately after switch-on with external sources and loads in every combination, as well as during normal operation.
3 Repeat 2, for switch-off after normal operation is reached.
4 Repeat 2, for switch-on followed by switch-off *before* normal operation is reached.
5 Repeat 3, for switch-on immediately after switch-off.
6 Check effect of excessive external input signals (if applicable).
7 Check effect of load short or open circuits and excessive voltages and currents generated by load (if applicable).

protection measures

Although each case must be examined individually, protection is usually achieved by inserting resistors to limit currents and adding diodes to limit voltages. In many cases a slightly different form of circuit will simplify the protection measures required.

CHAPTER 3

microcircuit usage

ALTHOUGH custom-built microcircuits are still uneconomical for small-quantity production, standard microcircuits, especially logic elements, are now quite inexpensive. Every designer is therefore likely to be associated with some assemblies that contain a few scattered microcircuits and others that contain only microcircuits.

When such elements first became available there was a general feeling that the designer could detail an entire system without circuit knowledge, simply by following the manufacturers' rules. It was believed by many engineers that the only future for the circuit specialist would be in the semiconductor manufacturers' design team generating new microcircuits.

Practical experience has shown this to be untrue for several reasons. The main one is that blind adherence to stated rules often leads to disaster because the manufacturer fails to envisage all possible applications. A designer may operate well within the stated limits, violate rules that should be stated but have been overlooked, and fail to obtain satisfactory operation. It is essential therefore that the designer understands in depth the functioning of the microcircuit.

Another reason is that in systems which require signal level changes that are unsuitable for microcircuits, e.g. 50 V excursions, the interface needs much more careful circuit design than usual because of the very wide tolerances which accompany the small voltage swings in the microcircuit.

It is also becoming clear that the manufacturers' own design departments will not be large enough to deal with the vast number of new designs which will be demanded by quantity users. This must lead to the design of microcircuits by circuit designers who are employed by the manufacturers' customers.

In the following sections several typical problems associated with both digital and linear microcircuits are discussed. It should be noted that some of these would not arise if the manufacturer had considered all conditions of use when writing his specification.

logic elements

In the standard TTL gate circuit the output stage is known as a totem-pole, four circuit elements being stacked vertically as shown in Fig. 3.1. A typical specification of output swing is $\leqslant 0{\cdot}4$ V when

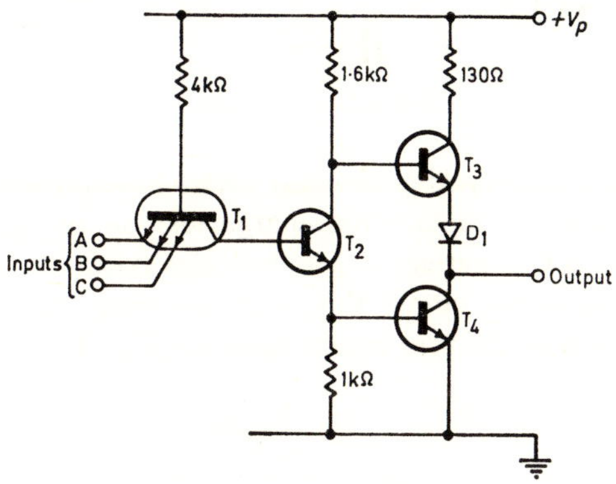

Fig. 3.1 TTL gate circuit

in logic 0, i.e. T_4 bottomed and T_3 off, and $\geqslant 2{\cdot}7$ V when in logic 1 using $V_p = 5$ V. These levels are guaranteed to be suitable for driving several other logic elements of the same species in parallel. The input circuit of this logic family in fact requires the logic 0 input to be $< 0{\cdot}8$ V and the logic 1 to be $> 2{\cdot}4$ V leaving 0·3 to 0·4 V of so-called noise immunity.

Supposing the unusual requirement arises that a.c. coupling has to be used between two such elements, the first of which is being driven between its two states with a 1/1 mark/space ratio. Two mistakes can be made if the circuit diagram of the microcircuit is not understood. The first is to capacity-couple using no other components, whereupon the output of the second circuit eventually reaches logic 0 and remains there. The reason is obvious by reference to the circuit: d.c. restoration occurs. On the first negative-going input on the capacitor the output of the second circuit is driven to logic 1 and during this time the coupling capacitor is charged through the input emitter. The input to the capacitor then goes positive, driving the second output to logic 0. When the input to the capacitor next descends, the voltage arriving at the input of the second circuit hardly turns on T_1 and this may or may not change the output state. Eventually C is fully charged so no further charge of state occurs. As in a normal d.c. restorer circuit, (Fig. 3.2) the current that flows

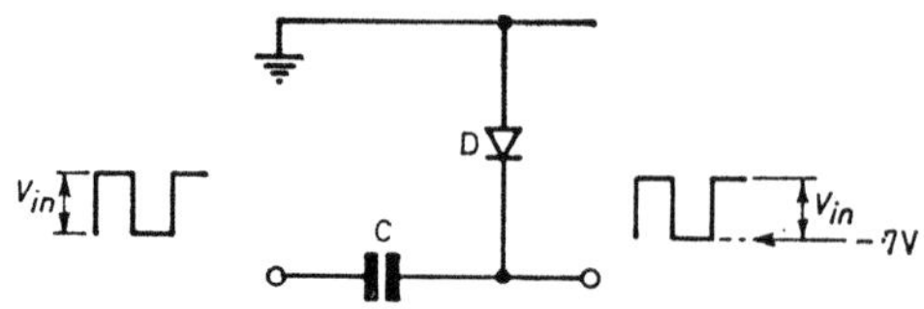

Fig. 3.2 D.C. restorer

on the negative half-cycle after many cycles have passed is only the amount required to replace the charge lost on the positive half-cycle, i.e. zero if C is perfect and no leakages exist.

By reference to the circuit this snag can partly be overcome by adding a resistor to the zero line as shown in Fig. 3.3.

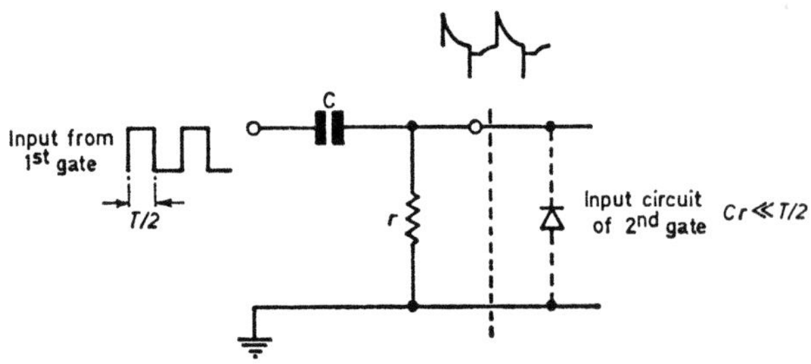

Fig. 3.3 Overcoming d.c. restoration (*but see text*)

Even with this modification the circuit is still unsatisfactory. If the input to C has been at logic 1 for long enough to allow C to charge fully, and the input then falls to logic 0, it attempts to move by at least 2·4 V. Since the output from C was at 0, the input of the second circuit is driven to $-2{\cdot}4$ V. If C is large, and the β of T_4 in the first circuit is high, enough charge can flow into the emitter of T_1 in the second circuit to cause its destruction. Some specifications do not allow the input to be taken negative under any circumstances.

Others allow this provided −1 V is not exceeded and that a series limiting resistance of 100 Ω is present. For these cases the circuit must therefore be modified as shown in Fig. 3.4.

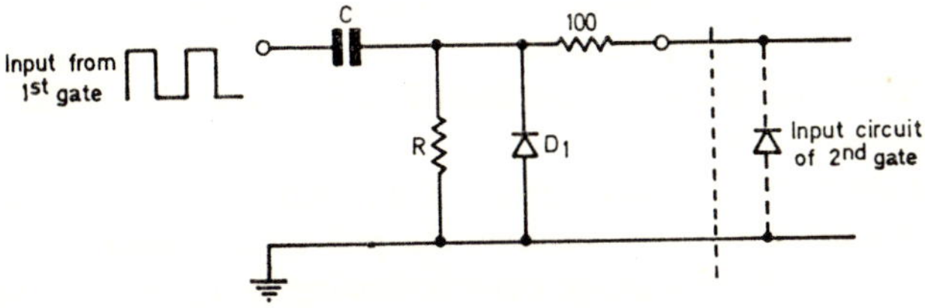

Fig. 3.4 Limiting negative current drive

Quite apart from safety aspects the circuits of Figs 3.3 and 3.4 are still uncertain in operation. Since the guaranteed swing at the input of C is only 2·4 V, the maximum positive voltage at the output of C is only +1·7 V due to the 0·7 V which is lost across D_1 when swinging negative. This is inadequate to switch the second circuit.

Examination of the totem-pole circuit (Fig. 3.1) shows that the small swing is due to the drops in the output diode and T_3 base-emitter voltage and losses in the base and collector resistors of T_3. Since T_4 is turned off when the output is at logic 1 it seems to follow that a resistor R from the supply to the output would ensure its reaching $+V_p$ (see Fig. 3.5). This is often used and R is referred to as a 'pull-up' resistor. Its use in conjunction with Fig. 3.4 appears to ensure correct operation.

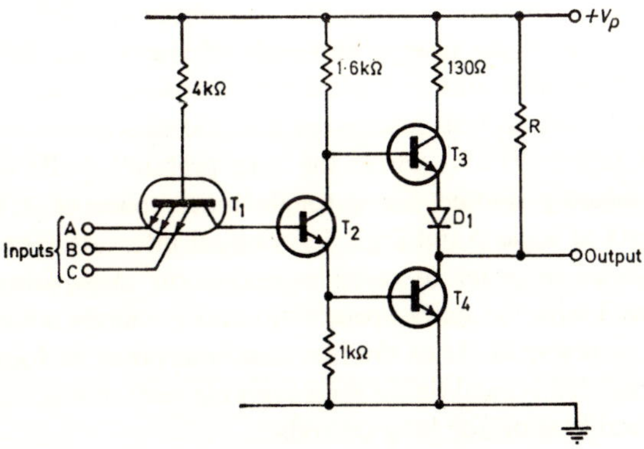

Fig. 3.5 Use of pull-up resistor R

There are still, however, two snags. The minimum value that R may be given is determined by the heaviest tolerable load when in logic 0. If at the p.r.f. used, the time constant RC_S (due to stray capacitance to earth C_S) is excessive, pull-up will not be achieved in time to switch the second circuit before the first circuit reverts to logic 0. For instance, when $R = 20$ kΩ, $C = 10$ pF, p.r.f. = 5 MHz, operation is uncertain (the pull-up certainly fails but the first circuit may deliver more than its guaranteed minimum so the second circuit could conceivably switch).

The other snag is that in spite of the existence of application notes by manufacturers recommending the pull-up technique, no indication is given of the leakage current flowing in T_4 when in logic 1. This turns out to have a very large value, e.g. >1 mA, in a significant number of microcircuits. Even when an extra check is demanded by the user, the manufacturer will rarely guarantee $<0{\cdot}5$ mA. Some units tested by the writer have been in excess of 10 mA and these units could not be rejected since they met every specification point.

The pull-up technique is therefore of use only if a logic 1 leakage specification is guaranteed. This fact greatly limits the use of micrologic elements as analogue driving amplifiers and makes more difficult the design of interface buffers between the logic and other circuits.

Another pitfall concerns power consumption. Most micrologic elements have a wide tolerance in the required supply current. This may be unimportant from the point of view of package dissipation provided the manufacturer guarantees that units will operate safely at a certain high temperature. When designing the power supply however, it is essential to know the maximum possible dissipation since in a large logic system hundreds of watts may be involved. Most manufacturers quote a typical dissipation rather than maximum and suggest (but do not guarantee) that the maximum could reach twice this figure. This is where the trap presents itself: the typical figure is usually quoted for a toggle mark/space of 1/1, i.e. the output rests at logic 0 and logic 1 for equal intervals. The factor of two suggested in order to reach a 'maximum' dissipation is quite unconnected with the mark/space ratio and is merely a rough guide to production spread. Thus the true maximum may be four or more times the quoted typical figure for a logic element that remains in the less favourable state for long periods.

Referring to Fig. 3.1 it is evident that a logic 0 output gives the higher dissipation even if the output load is negligible: yet application notes often recommend that unused gates of a complete package should have inputs joined to $+V_p$, thus giving maximum dissipation.

Naturally in a large logic system it is to be expected that not all gates will rest in one state so that the manufacturers' typical dissipation figure (doubled) may be a practical maximum. Bistables which often rest in a particular state for long periods do not affect the argument since dissipation is in these circuits similar in either state. Moreover it is only in a large system that power supply current becomes significant, and these facts justify to a large degree the 1/1 mark/space figure used. However, circumstances can arise where these assumptions no longer hold and it is advisable to assess each situation in detail before designing the power supply.

Another rating problem that affects the power supply design is the figure quoted for the maximum permissible supply voltage. A typical data sheet suggests that although for correct operation the supply voltage should be between 4·75 and 5·25 V, no damage is caused if the supply reaches 7 (or sometimes 8) V due to fault conditions.

The whole subject of power supply overvoltage protection has to be taken very seriously when a failure could destroy perhaps 1,000 microcircuits within milliseconds, and the 7 V limit makes this seem easy. Taken at face value this implies that the power supply could be a series stabilizer from a 7 or 8 V supply containing no special protection against failure of the series element. The supply would reach 7 or 8 V and although the logic may fail to work correctly no damage should result. Alternatively the supply could be derived from a higher voltage either by series-stabilization or by pulse-width modulation and switched off electronically if its output should exceed 7 V. The overvoltage sensing circuit could be simple as quite a wide tolerance of its trip level would be acceptable.

When other ratings are examined it soon becomes clear that although V_p might be 7 V, gate input terminals are all restricted to +5·5 V, thus restricting V_p to this figure if unused inputs are joined to the supply (as recommended by manufacturers!). Even if this practice is avoided, inputs driven from outputs of other microcircuits of similar type are still likely to be overdriven if these driving circuits have a supply that also reaches 7 V under fault conditions. Since, in general, all inputs within a logic system are driven in this manner, the supply limit must be taken as 5·5 V if pull-up resistors are used, or otherwise 6 V. This naturally makes overvoltage protection much more difficult but is the only safe criterion to follow if wholesale destruction is to be avoided when regulators fail.

One common belief which the circuit designer often has to dispel is the general impression (gained from much of the advertising material associated with microcircuits) that circuit performance is always better than in discrete circuits. This is sometimes true with regard to operating speed and thermal balance due to the proximity of components. Microcircuits often allow desirable but normally too-extravagant or complicated solutions to be adopted because in their internal design transistor elements are readily available in almost any quantity. When, however, a 'normal' circuit such as a one-shot, or binary, is made in this form, many of the features of good design practice are not available. The supply voltage, owing to breakdown-to-substrate is limited to a few volts. The circuits often have to 'lose' one or two base-emitter voltage drops and base currents, and voltage swings are therefore ill-defined. Resistor accuracy is usually no better than ± 20 per cent and in some processes *pnp* transistors have current gains of less than 5.

The use of one or two microcircuits of this kind in a system containing mainly discrete circuits is therefore likely to lead to disappointment, especially as the circuitry which drives in and out of these elements is difficult to design.

In considering which elements, if any, should be microcircuits, the specifications of the proposed elements should be regarded with even more than normal suspicion—they are likely to have very much wider tolerances than the discrete circuit they may replace. They will also have similar snags if the published circuit resembles a standard discrete circuit even though these may not be pointed out in the data sheet.

One concrete example is the DTL one-shot circuit given in Fig. 3.6. Its mode of operation is similar to that of the circuits given in Chapter 10, T_3 and T_7 being the principal transistors and $C_x R_x$ the timing components. Since it is difficult to make large capacitors in a reasonable area the maximum output pulse width is only a few microseconds with the internal components. By adding external capacitance and resistance this figure can be increased, the width being defined by $0{\cdot}5 C_x R_x$ approximately. (This is not the usual $0{\cdot}7 C_x R_x$ because V_{be} losses are significant in comparison with the supply voltage.)

Now, quite apart from the wide tolerance of this expression

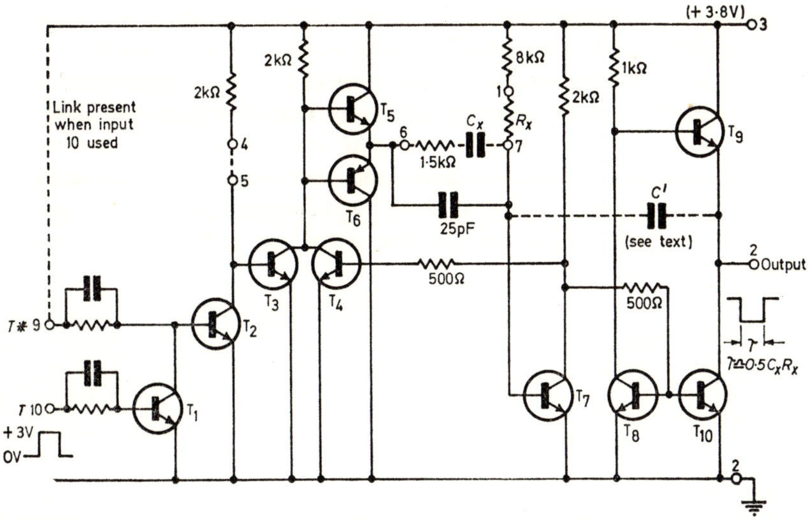

Fig. 3.6 Circuit of DTL one-shot multivibrator

(about ± 30 per cent due to small voltage swings and V_{be} losses), two pitfalls exist. Firstly, a very large external C_x, e.g. 10 μF, is likely to destroy T_7 when C_x swings positive at the end of the generated pulse. Secondly, an input pulse which lasts longer than the generated pulse holds on T_3 during the final recovery of C_x. This removes the regenerative loop round T_4, T_5 and T_7 and results in an output pulse trailing edge which is slower than specified on the data sheet if C_x is large.

Only after several unfortunate experiences did the data sheet suggest the need for a limiting resistor r in series with C_x when the latter exceeded 1,000 pF. The other point is still not mentioned but can be overcome by a 22 pF speed-up capacitor C' as shown—provided r is present. This gives regeneration round the output circuit.

Perusal of Chapter 10 will confirm that these failings are typical of fast-recovery one-shots and special measures must be taken if they are significant. The microcircuit is no better and this could be predicted by any circuit engineer familiar with the basic operating mechanism of the circuit.

The conclusion is that non-linear microcircuits are only as good as their basic design allows and that their shortcomings are usually predictable although not always mentioned in the data sheet. They are most advantageous when forming a complete logic system, requiring no special interface circuits, in which their guaranteed mutual compatibility is fully exploited. Their use in isolation is questionable: their tolerances in terms of operating time, voltage and current levels and their fragility require exact technique when interfaced with discrete circuits. Capacity-coupling between logic microcircuits is not recommended for several reasons given in the first part of this chapter.

CHAPTER 4

dynamic circuit faults

IN designing a circuit, the engineer may calculate d.c. conditions correctly for the static situation in which no alternating signals are present. It does not follow, however, that dynamic operation will be satisfactory, whether the signal is supplied from an external source or is generated by the circuit itself. The following examples illustrate some of the more common problems of dynamic operation.

clipper circuits

Perhaps the simplest example is the clipper circuit shown in Fig. 4.1a in which the input waveform is intended to be coupled faithfully

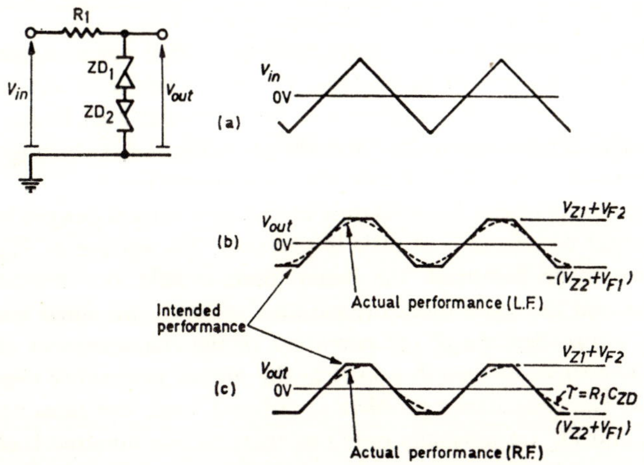

Fig. 4.1 (*a*) Clipper circuit and input waveform; (*b*) performance at low frequencies ($f \ll 1/R_1C_{ZD}$); (*c*) performance at high frequencies ($f \gg 1/R_1C_{ZD}$)

to the output until the clipping levels $\pm(V_z + V_f)$ are reached; the output is then supposed to limit at these levels.

The designer, naturally, must calculate the two clipping levels and ensure that initial Zener tolerance and temperature coefficient can be accepted. The dynamic situation, however, requires deeper consideration.

Firstly, Zener diodes have a 'knee' in the characteristic which is ideally rectangular as shown by the continuous line in Fig. 4.2; in

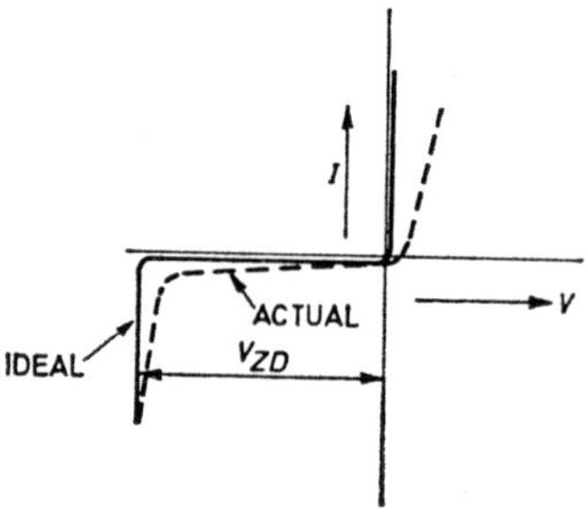

Fig. 4.2 Zener characteristic

practice the knee is more curved as represented by the dotted line. In the circuit of Fig. 4.1a, the Zener diodes will, then, begin to load R_1 appreciably long before the nominal Zener voltage (quoted at perhaps 5 or 10 mA) is reached. This results in severe distortion of the waveform as shown in the dotted curve in Fig. 4.1b.

Secondly, Zener diodes usually have quite high junction capacitance of several tens of picofarads. The performance of the circuit therefore depends on frequency and the output at high frequencies eventually degenerates to the form shown in Fig. 4.1c (which ignores the previous effect of the Zener knee, for simplicity).

This type of circuit is useful only in very non-critical designs where the output waveform is of little importance. The version of Fig. 4.3 is much to be preferred: the Zener diode is held at a reasonable current and the series diodes remain cut off until the signal reaches $\pm(V_z + V_f)$. Because of the curvature of the characteristics of D_1 and D_2 a slight distortion occurs at the output just before clipping occurs. The capacitance problem is reduced, since the capacitances of D_1 and D_2 are normally only one tenth to one hundredth of the Zener capacitance.

These comments apply equally well to clippers used for sine- and

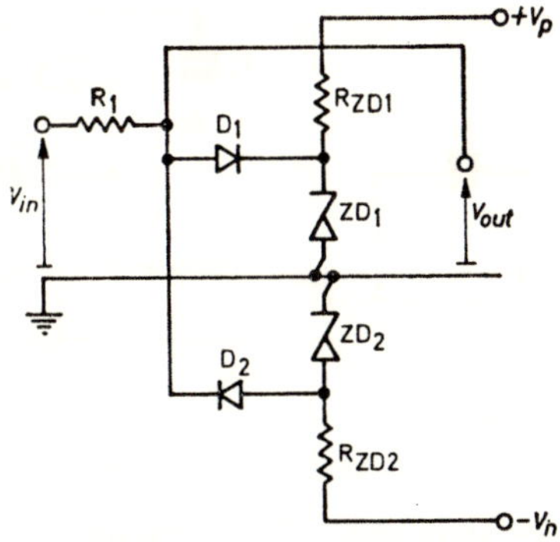

Fig. 4.3 Improved clipper circuit

square-wave signals. Although it is unfortunate that the preferred circuit requires d.c. supplies, it should be noted that when signal power, voltage level and frequency are suitable these supplies may be derived in the manner shown in Fig. 4.4. This arrangement is, of course, attractive only when no other d.c. lines can be found.

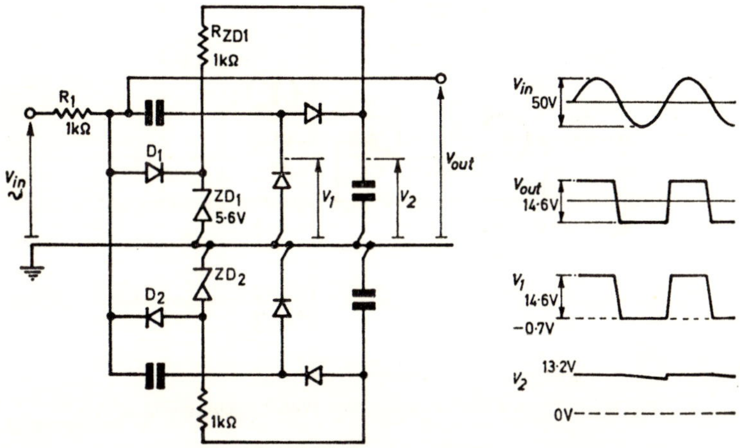

Fig. 4.4 Supplying d.c. Zener drive from input signal

integrator with limiting

The use of the active integrators shown in Fig. 4.5 is common in analogue computers, in servoloops and AGC control loops, since its transfer characteristic approximates closely to a true integral

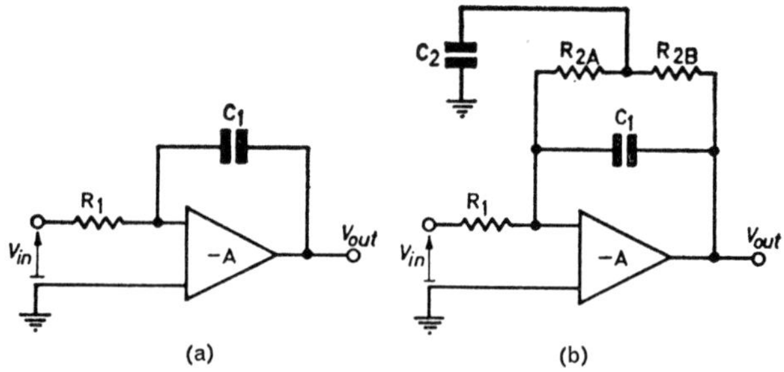

Fig. 4.5 Integrator: (*a*) basic circuit; (*b*) stabilized circuit

function $1/pT$ rather than the $1/(1 + pT)$ given by a passive single *RC* network. Its practical realization may use a microcircuit, e.g. μA 709, as the active element or this may instead be one of the discrete component circuits described in Chapter 13, as shown in Fig. 4.6.

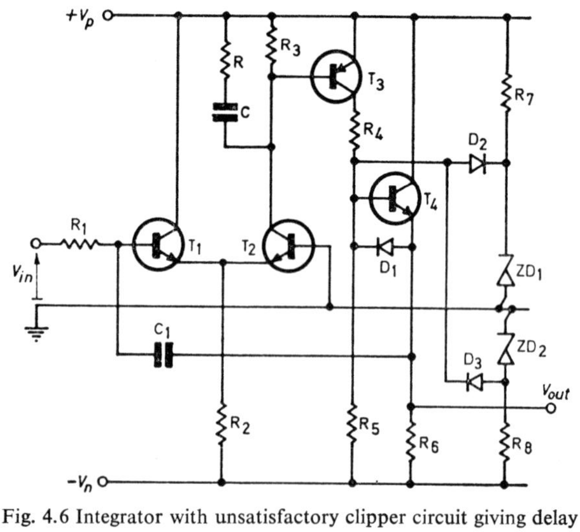

Fig. 4.6 Integrator with unsatisfactory clipper circuit giving delay

Such integrators are often used to obtain a reasonably accurate 90° phase shift for sinusoidal inputs. Two simple points in such usage are overlooked surprisingly often even by experienced designers. One is that, because of the absence of d.c. feedback, the circuit is unusable by itself; it must have some arrangement to hold the amplifier in its linear region such as a resistor of perhaps $100R_1$, in parallel with C_1. This degrades the 90° and the temptation is to split

R_2 into R_{2A} and R_{2B} and to decouple the junction to earth by C_2 (Fig. 4.5b.) Loop oscillation or a large discrepancy from 90° is the result unless a really careful analysis is performed to obtain the correct value for C_2. Often the integrator when used for sine waveforms is part of an over-all feedback system such as an oscillator; d.c. feedback can then be applied round the complete loop.

The other point is that the output of this circuit does *not* lag behind the input by an angle a little short of 90°, it *leads* by an angle slightly in excess of 90°. This is immediately obvious when examining the voltage appearing at the amplifier input. This voltage clearly lags almost 90° behind the input; the output is its inverse and therefore lags by almost 270°, which is equivalent in continuous sine-wave operation to a lead of just greater than 90°.

In use with non-sinusoidal inputs the circuit may drive a voltage-controlled variable attenuator, a voltage-controlled oscillator, a level discriminator such as a Schmitt trigger, or a servo-amplifier. In some of these applications it may be an essential requirement that the integrator output should not exceed a certain voltage. This could be for safety reasons (see Chapter 2) or to avoid for instance a condition where an oscillator ceases to function if overdriven.

The clipping circuit, similar to Fig. 4.3, is often used for this purpose and is connected as shown at the right-hand side of Fig. 4.6.

Note that the use of Zener diodes alone as in Fig. 4.1, although undesirable, would here not affect the output waveform nearly so badly as in that application. The integration would remain unaffected until the loop gain fell very markedly, i.e. when the Zener became conducting quite heavily.

With the circuit shown in Fig. 4.6 it is a simple matter to calculate the clipping level and to confirm that integration properties are unaffected until this level is reached (except during the turn on of D_2 and D_3 as in the simple clipper of Fig. 4.3).

A dynamic problem exists, however, in spite of the care taken to use a good clipper. If the input remains, at say, +4V for a long time then the lower clipping level is eventually reached at the output. An *immediate* fall of input to −4V then causes the output to rise linearly as soon as the input changes, until the upper clipping level is reached (Fig. 4.7). On the other hand, if the input remains at +4V for a very long time ($\gg R_1C_1$) before falling to −4V, T_1 base in the meantime itself rises to +4V because *the action of the limiter is to reduce the loop gain to zero* thus destroying the virtual earth effect. On taking

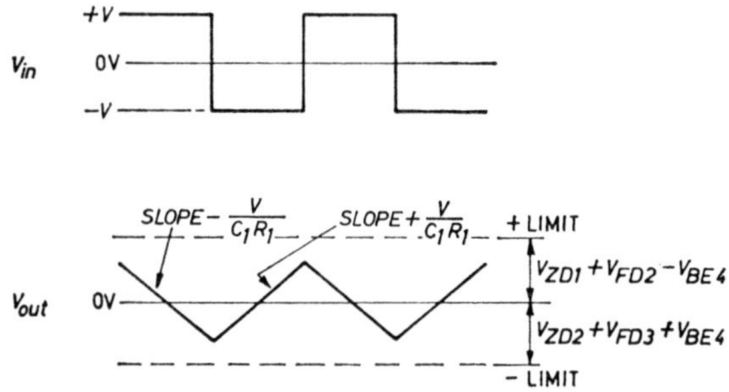

Fig. 4.7 Performance of circuit of Fig. 4.6 with no limiting

the input now to −4V, no change in output can occur (assuming a low output impedance from the amplifier) until T_1 base again reaches zero; this takes about $0{\cdot}7C_1R_1$ sec. Only after this elapsed time will the output begin to move, as shown in Fig. 4.8.

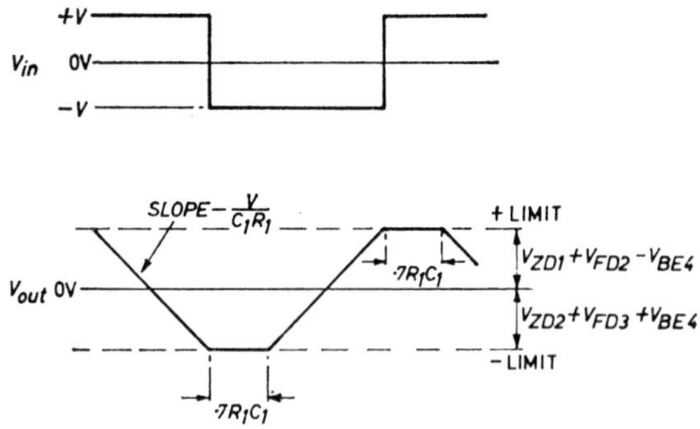

Fig. 4.8 Performance of circuit of Fig. 4.6 with limiting

This gives the effect, possibly disastrous in a feedback system, of an integration plus a time delay, the magnitude of the latter being dependent upon input amplitude.

The cure is to connect the clipper in such a way that the loop round the amplifier remains closed so that the amplifier input remains at earth potential. A rough method is shown in Fig. 4.9a where a simple double clipper is placed across C_1. This has, of course, the old snag that clipping begins half-heartedly, well below the nominal Zener level, and since this occurs outside the amplifier circuit, the output is affected directly. As the Zeners begin to conduct, some of the current

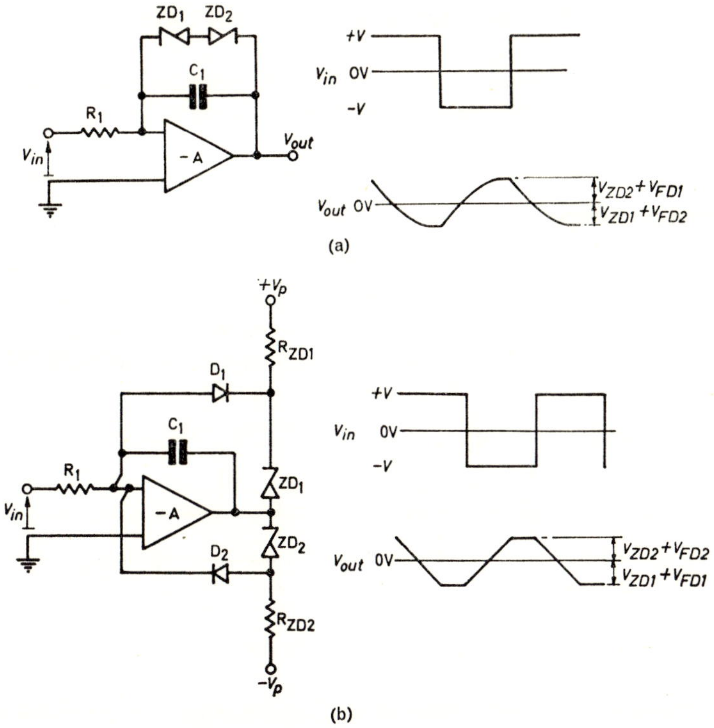

Fig. 4.9 (*a*) Bad clipper, connected in feedback loop gives soft clipping but no delay; (*b*) good clipper in feedback loop gives correct clipping without delay

from R_1 which had been charging C_1 linearly is now diverted and the rate of voltage change on C_1 falls, giving a curved output. When the output finally reaches $-(V_z + V_f)$ it remains at this level and *because there is feedback via* ZD_1, ZD_2, the amplifier input remains near earth. On reversing the input potential, the output therefore changes immediately it begins its ascending ramp.

In order to avoid the curved characteristic it is necessary to use an arrangement whereby the Zener diodes are permanently conducting as in Fig. 4.3. The configuration must be so chosen that the standing Zener currents have a suitable path which have no appreciable effect on the charging process. Fig. 4.9b shows a suitable method in which the amplifier output accepts the two Zener currents. A design point, if using an amplifier form similar to Fig. 4.6, is to ensure that R_6 is sufficiently small. The danger is that T_4 may cut off at maximum negative output if the difference between Zener currents from R_{ZD_1} and R_{ZD_2} exceeds the current provided by R_6.

The form of circuit shown in Fig. 4.9a provides clean limiting action with negligible delay even after prolonged excessive input.

chopper circuits

To minimize drift when amplifying d.c. signals, chopper techniques are often used. In its simplest form the chopper can consist of a single transistor (either a bipolar or field effect type) which is driven on and off, thus switching the d.c. input signal intermittently to earth (Fig. 4.10).

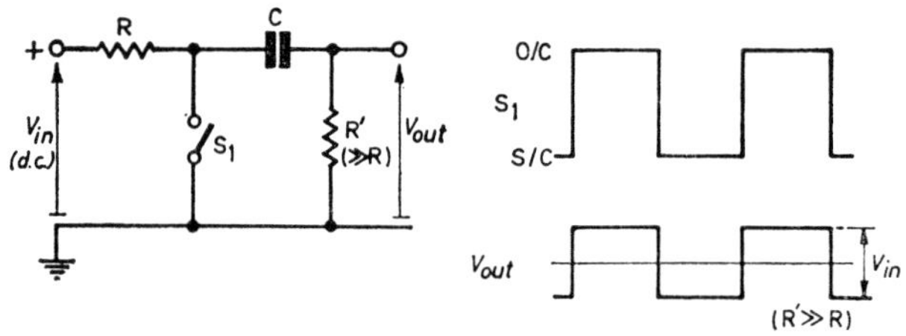

Fig. 4.10 Simple chopper

The signal now contains an alternating component proportional in amplitude to the d.c. input and can therefore be a.c. coupled and amplified. The alternating output, which is finally rectified, is not affected by d.c. zero drift in the amplifier.

The realization of this technique is described at length in Chapter 9 of EDH.

A problem arises when coupling the S_1 signal to the amplifier by means of capacitor C. The function of C is to remove the d.c. component so that if the mark/space of switching of S_1 is 1/1, the signal at the amplifier input is disposed symmetrically around 0V. In designing the amplifier consideration has to be given to its input impedance R', ideally much greater than R. In many cases this is a difficult condition to meet and it appears satisfactory to make it a known value, comparable with R, and simply allow for the resulting attenuation in the gain calculations.

This seems such a straightforward concept that it may come as a shock to find that with $R' = R$, V_{out} is reduced to $\frac{2}{3}V_m$ pk-pk, *not* to $\frac{1}{2}V_m$ pk-pk!

The automatic assumption that there is a factor-of-two attenuation is a mistake made by the beginner, who ignores the capacitor; not by the slightly more experienced engineer who sees the capacitor and cannot guess its effect (and therefore calculates the result step by step). The same mistake is again made by the more senior man who notices the capacitor and thinks it may be safely ignored since sharp transitions are involved; but not by the long suffering engineer who has learnt to be suspicious of any circuit in which a capacitor has different charging paths during one cycle of operation.

In any circuit involving capacitors many operating cycles may occur after switch-on until an equilibrium condition is reached. This state can be recognized by the fact that all waveforms are identical from one cycle to the next. It follows that the net charge received or lost by any capacitor in the circuit must now be zero during any one cycle. In some circuits, e.g. Fig. 4.10, this will imply that the capacitor carries a direct charge when equilibrium is reached.

To calculate the behaviour of a *CR* network under these conditions it is sufficient to write an equation expressing the above condition.

Using Fig. 4.11 as a typical example, assume C is extremely large so that the alternating voltage across its terminals is negligible. When

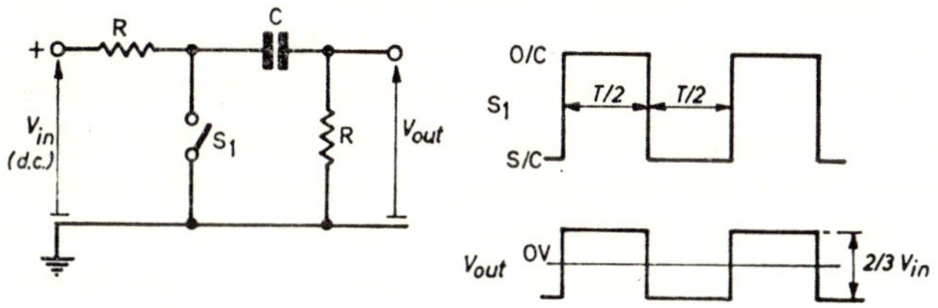

Fig. 4.11 Loaded chopper

equilibrium is reached C therefore carries an unknown voltage V_c and behaves rather like a battery of this voltage.

To calculate V_c, the two conditions of S_1 are examined and the net charge received is equated to zero. When S_1 is closed, C loses charge through R; since *C* is extremely large, $CR \gg T/2$ and the amount of actual loss of V_c is negligible, i.e. no a.c. component across C, so that the discharge current is virtually constant at a value V_c/R. The loss of charge with S_1 closed is therefore $\frac{1}{2}T(V_c/R)$. When S_1 is open C gains charge from V_{in}; the current is again constant throughout the half cycle as is $(V_{in} - V_c)/2R$, assuming that the positive plate of C is joined to R, giving a charge gain of $\frac{1}{2}T[(V_{in} - V_c)/2R]$. The net gain or loss must be zero, so that

$$\tfrac{1}{2}T(V_c/R) = \tfrac{1}{2}T[(V_{in} - V_c)/2R]$$

This gives $V_c = V_{in}/3$ and is positive, i.e. is confirmed to be in the assumed direction.

As in all such analyses, knowledge of the capacitor charge enables the other waveform levels to be deduced easily. A simple procedure here is to observe that with S_1 closed, V_{out} falls to $-V_c$, i.e. $-V_{in}/3$. Since the output is capacity coupled and the mark/space is 1/1, the levels must be symmetrical and so when S_1 opens V_{out} must rise to $+V_{in}/3$.

Note especially that the charge CV_c does not appear at switch-on but requires a time, determined by CR, to reach this equilibrium state. This implies that over-cautious design to ensure negligible signal loss by using unnecessarily large C means that the 'warm-up' time is extended – during which time the amplifier receives a 'd.c.' signal, probably causing limiting. Also note that, although it may seem pointless having now analysed the circuit, the first instinct of many designers to avoid building up such a charge is to increase C: this merely lengthens the agonizing waiting period until the conditions reappear (cf. Chapter 1, pages 19–20).

To summarize, it pays to be suspicious whenever a capacitor possesses different charging paths during any one cycle of operation. The equilibrium state should be calculated as shown above and it must be remembered that this condition is reached only after several cycles. This pre-equilibrium sequence can be calculated by simply assessing the gain of charge each half cycle. (For example, when S_1 first opens C gains $\tfrac{1}{2}T(V_{in}/2R)$ and therefore reaches a voltage $V_{c1} = \tfrac{1}{2}T(V_{in}/2CR)$; it then loses $\tfrac{1}{2}T(V_{c1}/2R)$, so falls by $\tfrac{1}{2}T(V_{c1}/RC)$ and so on.)

The main problem in all the above circuits lies not in the technical difficulties viewed objectively but in acquiring an awareness that a difficulty exists; the solution is then close at hand.

CHAPTER 5

component choice

ONE common reason for the failure of a new circuit design is the lack of attention given by the designer to the choice of components and the geometry of their interconnection. This is a natural occurence because the challenging part of the design has been completed and the practical details seem trivial.

It is assumed that the designer is familiar with the normal properties of components and is therefore capable of making a sensible choice. He will also have access to manufacturers' catalogues which usually indicate the tolerance limits associated with the various components.

The following notes, based on actual cases brought to the writer's notice, are devoted to component or layout features which are not to be found (except perhaps in the small print!) in manufacturers' catalogues. For a much more searching survey on similar lines the reader is recommended to refer to *Reliability of Electronic Components* by C. E. Jowett, Iliffe Books, 1966.

resistors

In deciding on the tolerance it is essential to take account of temperature coefficient and drift with time (ageing). Some resistors are 'triple rated', implying that the specified tolerances are met even when these are all considered. This information alone is not usually sufficient; the designer is interested in the separate contributions to changes in value. It may be valuable to know, for example, whether the temperature coefficient is positive or negative; often the specification allows either polarity.

Clearly the expected change in value due to temperature is related to the actual temperature rise of the component. If this is itself dissipating appreciable power, or is situated close to a radiating source of heat, its temperature may be considerably greater than the ambient temperature of the whole equipment. It is true that this is unlikely to alter the change of value when the ambient changes since the component is a certain number of degrees higher than the ambient at all times. However there will be a considerable change of component temperature during the first few minutes after switching on.

Power rating of resistors can be misleading, being dependent upon the method of mounting. Some rely only on heat conduction along the leads; others rely on convection of free air flowing past the component body and others on radiation. It is therefore, unwise to mount a wire-wound resistor directly on a circuit board, cover it in protective varnish and expect its normal free-air rating to be maintained.

Potentiometers present a special rating problem. The wire or track has a maximum, safe current-density rating which sets a maximum power rating for the potentiometer assuming this is dissipated along the whole track. If the amount of track being used is very small, the control being set near one end of its travel, the amount of allowable current is still determined by the track or wire. The total permissible power is therefore greatly reduced. The relationship is not quite so simple as the above would suggest due to heat-sinking. A safe rule is to keep the current level below the value which corresponds to the maximum dissipation for the whole track, even if only a small part of the track is used.

When direct current is applied to a resistor several side effects occur apart from the resulting power dissipation. Most types of resistor become non-linear to a degree which is insignificant in normal usage but which causes puzzling effects in low-distortion circuits.

Another result of direct current is an increase in the low-frequency noise generated in the resistor. This noise is of the type which rises as frequency falls and is usually known as '$1/f$' noise. Metal film and wire-wound resistors are free from the above effects; other types must be avoided if ultra-low distortion ($<0{\cdot}01$ per cent) is sought or if low noise is important. Most manufacturers will quote a 'noise versus applied-d.c.' formula.

In most industrial or military equipment, potentiometers have wire-wound tracks. In many other applications, especially audio, carbon tracks are preferred; though of shorter life, they are inexpensive and are free from the sudden, though small, jumps in resistance exhibited by the wire-wound track. This gives smoother operation when the track is still new, though noise often becomes troublesome after several thousand operations. The main rule when using carbon potentiometers is to avoid passing any appreciable direct current through the track which invariably leads to excessive noise similar to that produced by a carbon microphone.

Without doubt the most popular resistor in industrial and military use is the metal oxide type which is useful in almost every application outside the precision category, the total excursion due to all effects being about 7 per cent. The cheapest carbon types, though having initial tolerances of typically ± 10 per cent, have total excursions of around ± 25 per cent and have only limited application. In spite of its good features, at least one metal oxide type in common use suffers badly from a thermocouple effect. This is readily demonstrated by connecting such a resistor of several kilohms directly across a 20,000 Ω/V voltmeter and applying heat, e.g. from a soldering iron, to one resistor lead. The type in question produces a good 10 mV, easily read on the 50 μA range (equivalent to 125 mV f.s.d.). In low level d.c. amplifiers the resistors must therefore be chosen with care, and the thermal problem must be regarded as quite different from the temperature coefficient of resistance. The latter depends on the track, the former on the method of end connection inside the resistors.

Low value resistors of a few ohms or less are usually specified as wire wound due to a general lack of reasonably toleranced resistors of other types below 10 Ω. Before making this decision, however, it is always worth considering the use of a few carbon or metal oxide types in parallel. Surprisingly, the smallest wattage wire-wound resistors of any repute cost around five times the price of a metal oxide resistor.

For low inductance, spiral-wound and simple wire-wound types should be avoided; as a measure of the importance of this, note that a small 33 Ω wire-wound resistor may measure 35 Ω impedance at only 10 kHz. Stray capacitance appears in distributed form to earth when a resistor is mounted in a clip, or close to a metal plate, and also across the resistor terminals. In circuit design it is advisable to assume up to 0·5 pF for the latter, in the absence of more reliable information.

When a set of resistors, closely matched in temperature coefficient is required, as in a digital-to-analogue converter or an attenuator network the least expensive solution is often a thin film network. When several resistors are involved, a total quantity of only 20 networks usually justifies the cost. Moreover the over-all performance is generally superior to that obtained from precision resistors owing to the consistency with which the coefficients match in any one network and the ease of maintaining constant temperature across the substrate.

In professional equipment it is normally considered bad practice to use potentiometers incorporating a switch at one end of the travel. Although in common use in domestic equipment, e.g. for 'volume and on/off', no switch design of this type so far seems to meet industrial or military standards. The arrangement whereby the potentiometer spindle extends well beyond the control operating a separate switch by external linkage is, however, acceptable. One final point concerning variable resistors: Fig. 5.1a shows the obvious method of connection of a potentiometer when used as a variable resistor, terminal C being unused. The mode of connection preferably employed is shown in Fig. 5.1b, the slider being joined to

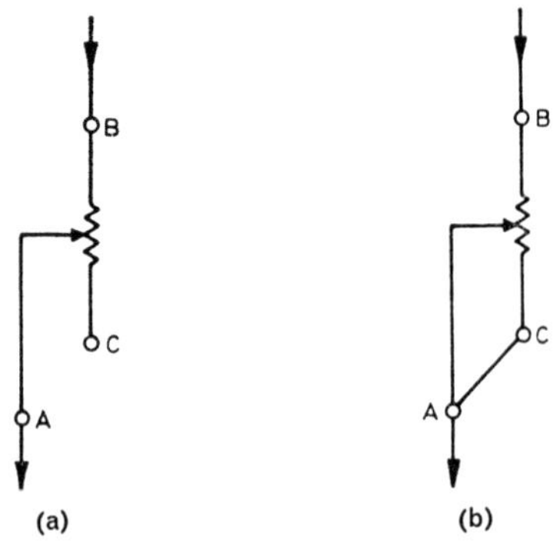

Fig. 5.1 (*a*) Variable resistor; (*b*) preferred connection for variable resistor

terminal C. The latter connection has the advantage that, should the slider lift off the track momentarily during adjustment (which often happens with a worn carbon potentiometer), the resistance between A and B jumps only to the full value of the potentiometer, not to an infinite resistance. This can be particularly important in a high-voltage resistor chain as used in defining electrode potentials on a cathode ray tube. In this application, a potentiometer that open-circuits the chain may result in an arc between slider and track, rapidly burning out the potentiometer.

capacitors

Although similar tolerance problems arise when choosing capacitors, the main difficulties are concerned with dynamic properties. Capacitors that are physically large tend to have appreciable series inductance thus degrading high-frequency circuit performance. For example, a 1 μF paper capacitor is likely to be less useful in decoupling a 10 MHz signal than a 0·001 μF ceramic type. The common practice of using two capacitors, with different values, in parallel, one for medium and one for high frequencies is of only limited application; there is always a high frequency at which the inductance of the large capacitor resonates with the two capacitors in series.

The effect of series resistance (poor power factor) shows up at quite low frequencies in sinusoidal applications where phase change is critical and in ramp generators which must not exhibit initial steps or over-all non-linearity. It is useful to convert the quoted power factor of a capacitor into the equivalent series resistance before the decision is made to use a particular dielectric. The effect on a circuit then becomes much easier to envisage and a surprising number of designs are in trouble on this point.

Where a capacitor has to charge slowly and then discharge rapidly as in an oscilloscope time base, the type must be chosen with special care. Some circuits of this kind rely (inadvertently) on the presence of series inductance to re-set the discharge circuit. This is a dangerous situation since a subsequent improvement in the capacitor, such as thicker internal connection pads could cause circuit failure. The well-known complementary discharging circuit of Fig. 12.1 can only too easily be designed in this way. Naturally, if this mode of operation is desired the inductor should be deliberately incorporated as an extra component (see Chapter 12).

Polyester film dielectrics exhibit weird characteristics in circuits like Fig. 12.1 (even when correctly designed). The effect is to produce severe distortion of the charging curve in the region immediately following the discharge. This appears to be caused by dielectric storage, often seen in large capacitors which can recharge themselves after repeated crowbar discharges.

Mechanical effects may be important in equipment likely to be subjected to vibration. Some plastic dielectrics act as piezoelectric transducers, generating an e.m.f. and changing considerably in capacitance when slightly compressed. This can lead to microphonic effects if used in sensitive parts of audio amplifiers.

When small physical size is important, ceramic capacitors of very high permittivity dielectric are often used for values up to 0·1 μF. Although adequate for decoupling purposes, having lower inductance than metallized paper types, they have two major disadvantages. One is the high leakage when operated near to the rated voltage—this is typically tens of microamperes at 30 V. Another is the temperature coefficient, often as high as 2%/deg C which may give, in military or industrial equipment, an over-all variation of 2:1 over the temperature range.

Electrolytic capacitors tend to be regarded as necessary evils. Great care must be taken in establishing operating conditions well within specified limits. In the case of many other components, including transistors, it is usually quite safe to operate at the stated maximum voltage. Electrolytics, however, are very closely specified with virtually no safety margin; moreover, operation just within the voltage ratings gives very large leakage compared with operation at 2/3 the quoted voltage.

Although it seems an easy solution to operate at a fraction of the rating this gives very odd results peculiar to electrolytic capacitors. If a capacitor which is labelled, for example, 100 μF/12 V, is operated at 6 V to improve leakage and operating life, the electrochemically formed dielectric gradually becomes thinner. Eventually, the capacitor becomes fit for no more than 6 V and its capacity increases to perhaps 200 μF. In a design intended to operate for several hundred hours the designer can only assume that electrolytic capacitors will leak as much as the manufacturers figure at full working voltage.

The biggest rating problem appears when dealing with reservoir capacitors for power supplies. This is dealt with in detail in Chapter 7 and is concerned with ripple current. Unfortunately, many manufacturers are unwilling to quote firm ripple current figures which are sometimes limited by dielectric heating and sometimes by the connecting pads and leads.

Tantalum electrolytic capacitors have been used in military equipment for many years. Only recently is their use becoming widespread, owing to the advent of inexpensive solid tantalum types. The reliability advantage over conventional electrolytics is no doubt well known to the reader. Less known, however, is the limitation in surge current capability of tantalum types, requiring a circuit series resistance of at least 0·3 Ω/μF.V. This implies that they cannot be used for power supply reservoir work unless load currents are very low. Wet tantalum types are still available; some of these have the considerable disadvantage that rotation of the component through 90° can result in considerable capacitance change due to movement of the electrolyte.

Reversible electrolytics may be of the specially made double-anode type or merely two normal types connected in series (positive-to-positive or negative-to-negative) and repackaged as one device. The theory of operation, especially near zero applied voltage, is very complicated and little understood even by the manufacturers. Whenever possible their use should be avoided, bearing in mind that tantalum types will operate happily when reverse biased by up to 1 V.

inductors

Until recently, the use of inductors has been avoided as much as possible by circuit designers. This has been due to the lack of availability of well specified inductors so that most companies have been obliged to design the winding and specify the core, to be manufactured outside. Some have undertaken production also, with the attendant problems of encapsulation and testing.

Most electronic equipment companies are not skilled in component manufacture, and the resulting practical difficulties have caused most designers to use only capacitors and resistors. This closes the vicious circle which makes entry into the inductor market unattractive to component manufacturers.

Quite recently, however, high quality inductors have become available from a few reputable manufacturers in a resistor-like shape. Values extend to tens of millihenries and prices are comparable with capacitors. Inductors may therefore be treated as 'normal' components and their exploitation often produces economic designs in pulse-generating circuits.

Transformers still require special winding because of the enormous variety required. The advent of ferrite toroids, at prices comparable with capacitors, has again produced changes in attitude. For development work these cores may usually be wound by hand within about an hour at a low over-all cost.

Large quantities are readily produced by specialist companies with toroidal winding equipment; although these windings present some difficulties, over-all cost is low in large-scale production. Toroidal cores of ferrite or strip are much cheaper than laminated stampings which require re-annealling after stamping manufacture.

relays

Relays, like electrolytic capacitors, are generally avoided by the designer. In many applications they have distinct advantages over semiconductor switches, for instance input/output isolation, the ability to withstand several hundred per cent short-term overload (both coil and contacts), better on-and-off resistance, no signal given to the drive circuit when the output current changes, and the ease of obtaining control of several isolated outputs.

When relays are used to switch heavy currents in inductive circuits it is advisable to limit the induced voltage which may appear across the contacts. This could be done by a parallel capacitor which limits the induced voltage caused by turning off a current I in an inductance L to $I\sqrt{L/C}$. However, if the contacts are closed when C is charged, a very large current flows, limited only by the series resistance of the capacitor, the relay contacts and the wiring. A resistor must therefore be connected in series with the capacitor to limit the current to a safe value; this modifies the turn-off voltage which must be recalculated.

Unsealed relays suffer from contamination by dirt which occasionally lodges between contacts. If such a relay is used it is important to use twin contacts and to arrange the mounting to make the contact blades vertical. This helps to ensure that swarf and dirt will not settle on the contacts.

Sealed relays should in theory be free from this problem; however, dirt and swarf are often already in the relay case, become sealed in for life and cannot be seen from outside! It is therefore even more important than usual to buy from a reputable manufacturer.

CHAPTER 6

system defects

SEVERAL volumes could be written on this subject but the intention here is to restrict the scope of the chapter to system defects that may easily be mistaken for circuit misbehaviour. Much time can be spent trying to locate a fault in circuit detail when examination of the system as a whole indicates the need for a completely different mode of operation. The following is devoted to such defects rather than the errors caused when a system, correct in principle, is too loosely defined, leading to an over-all specification not being met.

Most system errors are characterized by some irreversible chain of events which lead to an unwanted situation from which recovery is impossible.

For example, in a steel-rolling mill the sheet of steel is moving horizontally from the rollers and has to be cut to length. It is easy to design a transducer which detects when the end of the sheet has reached a certain point. This can take the form of a photoelectric detector and lamp so that the light signal disappears when the sheet comes between. Alternatively it may be a magnetic or capacitive proximity device. In either case, a pulse can be obtained from the transducer which triggers a one-shot circuit. The one-shot circuit can turn off the power to the rollers and activate the cutting device; after a time determined by the one-shot components, roller power is returned and the cutting device de-energized. The next sheet is cut in the same way.

This may seem at first a reasonable solution and careful circuit design would result in a workable system. In fact the system is so dangerous it could hardly be more badly conceived. Imagine that one pulse from the transducer fails to trigger the one-shot. The rolling would continue until turned off manually since after missing one pulse and failing to stop and cut, no further pulses would ever be produced. No apology by the designer, however phrased, would compensate for the resulting damage to equipment and personnel!

In a case like this where a missed pulse would cause untold damage, it is futile to put the onus on circuit design to ensure that a pulse cannot be missed. It must always be assumed in system work that sooner or later there is certain to be a missed pulse. Many outside influences can combine to cause this, e.g. a lightning flash, mains surge, a blown fuse, RF interference, switch-on conditions, etc.

A solution to the example quoted is to produce a change of level (d.c. or a.c.) when the sheet reaches the correct position. This level causes the rolling to stop and the cutter to operate. When the now-severed sheet is removed the level reverts to its original value and causes rolling to begin.

This is a simplified statement of the problem and the solution needs some refinement but at least the principle is sound; rolling can never take place when the sheet is in its cutting position.

Not many applications are as critical as the above but keeping in mind the assumption that pulses will inevitably be missed by triggering devices helps to avoid comparable situations. In some equipment, such a system fault is unimportant provided a resetting switch is provided for the operator.

Another premise of a similar kind is that a normal bistable circuit (in either discrete component or integrated circuit form) cannot be relied upon to remain in one state unless held there by a definite signal.

Unless special measures are taken, e.g. Fig. 6.1, a simple test soon shows this to be true. It is only necessary to monitor the state of the circuit as shown, in a manner which should not cause triggering due to pick-up on the monitor lead, and to switch on the supplies. Whether the supplies are mains-derived or from a battery, decoupled or not; whether the circuit is screened by one or more metal cans; whether housed for the test in a screened room, the state of the circuit will inevitably change if left for a few hours. The addition of C_1 and C_2 improves the situation, and other configurations such as the emitter-coupled binary are less prone to spurious triggering. These variations are, however, only changes of degree and a finite risk is still present even though the time interval between random triggerings may be much longer.

In using binary circuits as long-term storage elements this risk must be recognized and periodic re-setting or correction used. Finding a

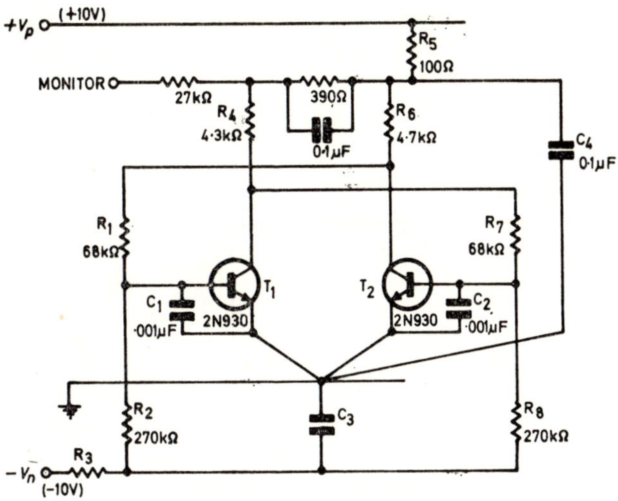

Fig. 6.1 Binary circuit

binary in the wrong state must not spell disaster to the system and this possibility must be regarded in a similar light to the lost pulse referred to above.

The situations described above imply that logic systems should in general be directly coupled. A level change should be used rather than a pulse generated from the edge of a transition. When this cannot be done, definite steps must be taken to ensure that a lock-up condition is impossible.

triangle generator with a.c. logic coupling

Another example, frequently seen in practice, where an undesirable state can be reached, is shown in Fig. 6.2a. The idea is to generate the triangle and square wave indicated in Fig. 6.2b. The Schmitt trigger circuit is a.c. coupled to one base of the binary and therefore drives the binary to one or other state according to the direction of the Schmitt output. The idea is that when the binary output is $+V$ the integrator output falls linearly (since C is charged by a constant current V/R). On reaching the lower trigger level $(-V_1)$ the Schmitt produces a negative swing which is a.c. coupled to the binary, changing its state. The binary now has an output of $-V$ and the integrator ramps upwards until reaching the upper trigger level $(+V_1)$. The Schmitt now produces a positive swing which again reverses the state of the binary.

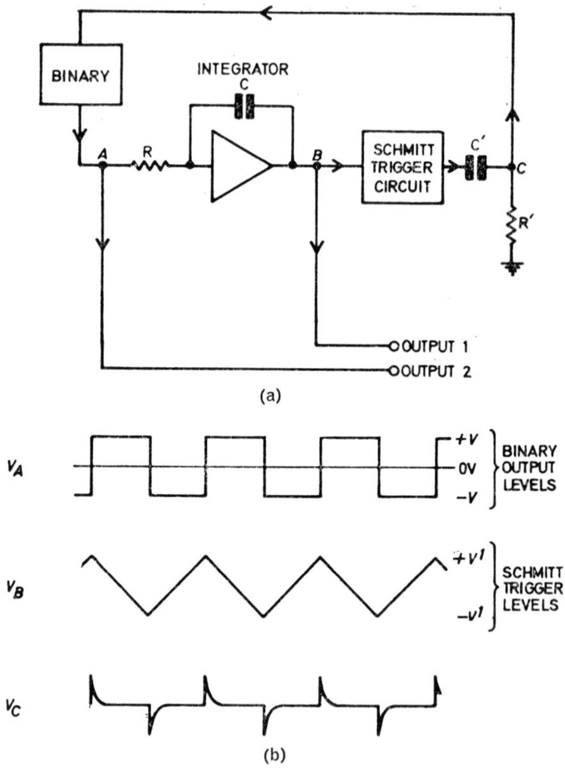

Fig. 6.2 (*a*) Triangle/square-wave generator; (*b*) waveforms for (*a*); *but see text*

The experienced engineer would immediately be suspicious of the a.c. coupling; closer investigation shows that he would be justified. This system has similar characteristics to the disastrous rolling-mill control already discussed.

Supposing the switch-on condition of the circuit is such that the binary output is at $+V$ and that the Schmitt contains a decoupling arrangement making it ignorant of input signals for a time longer than V_1CR/V. As the integrator output passes $-V_1$ there will be no output from the Schmitt and the binary will remain in its existing state, delivering $+V$ to the integrator. The integrator output continues to fall until its output circuit limits and there the circuit rests.

Switching on and off the supply lines may cause the intended conditions to come about. A more certain method is to short-circuit the timing capacitor C, bringing the integrator output to zero, which lies between the trigger levels V_1 and $-V_1$. On releasing the short-circuit the integrator output moves until the Schmitt has another opportunity to switch the binary.

The real cure is of course to eliminate the a.c. coupling $C'R'$ so that at all times after the Schmitt trigger levels are exceeded, the binary will be pulled into the required state. It is then easy to see that the binary is not required at all, provided the Schmitt output levels are $+V$ and $-V$. This is difficult to achieve using the standard two-transistor Schmitt circuit if V and V_1 are determined by external requirements (since they fix the output swings from the square and triangular outputs). Reference to Chapter 11, however, shows how a three-transistor Schmitt circuit will readily do this and Fig. 6.3 shows a directly coupled version of Fig. 6.2a using this method with the aid of an inverting transistor for correct over-all feedback sense.

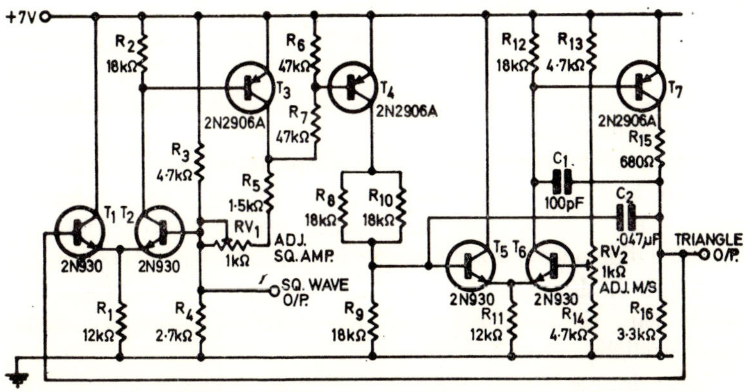

Fig. 6.3 Practical triangle/square-wave generator

The excursion of T_2 base is about 2 V pk-pk, resulting in square and triangular outputs of this amplitude. The integrator uses the three-transistor amplifier described in Chapter 13, R_{15} being added to limit the excursion presented to T_1 base to a safe level.

This d.c. coupled loop avoids the embarrassment caused by missed pulses but brings a problem of its own which is likely to trap even the more experienced designer. This is the tempting idea to drive T_4 base from T_1 collector (removing R_7). Loop phasing is preserved, yet the circuit fails to operate, settling with all conductors in mid-conduction. The reason is that after T_1 base potential exceeds the upper trigger level, T_2 and T_3 turn off and T_4 drive remains under the control of T_1 base potential. This forms a stable negative feedback loop.

Such a condition is a probability in many similar loops in which the over-all d.c. feedback is negative. To ensure the intended mode of oscillation, these loops must have *time-lag and voltage feedback* as in Fig. 6.3.

heterodyne loops

One of the problems in signal generation is the maintenance of constant signal amplitude over a wide range of frequency. Most directly-tuned oscillators have a range of only about 10 : 1 in frequency before their amplitude changes exceed a few per cent. If the waveform does not have to be sinusoidal there are many constant amplitude circuits with extremely wide frequency coverage (see, for example, Chapter 11 of EDH for a 40 : 1 variation with a voltage controlled oscillator).

For sinusoidal generation, it is therefore attractive to use a heterodyne (or beat) system. If a fixed oscillator of, say, 1 MHz is heterodyned with a variable oscillator tuned from 1·01 to 2 MHz the resulting difference frequency varies from 10 kHz to 1 MHz, a range of 100 : 1. If the mixer is driven correctly, the variable frequency signal being the switching input, the amplitude variations can be made negligible.

This principle is often used for wide range audio-oscillators. The snag is that a small change in the frequency of one or other oscillator causes a much larger percentage change in the output frequency – the very feature being exploited. In the above example a change of 1 per cent in the fixed oscillator implies a 10 kHz change in the output frequency, representing 100 per cent at one end of the range.

The usual 'cure' is to provide a setting-up control to adjust the tuning of one oscillator; since adjustment is required frequently the use of the technique in this form is restricted to test gear in which the control can be mounted on the instrument panel. It is then adjusted several times in a day. Elementary design precautions are usually taken to make the oscillators tend to drift by the same amount in the same direction, e.g. by using similar circuits and mounting arrangements. In practice, elaborate electrical isolation has also to be provided between the oscillators since, as the variable oscillator is tuned closely to the fixed oscillator in frequency, there is a strong tendency for the two to 'lock' to the same frequency.

The advantages of beat systems are so attractive that attempts are often made to overcome the obvious snags: inaccurate output frequency; tendency to lock together, giving no output beat. A more subtle problem is the distortion that occurs as the beat frequency becomes small (see *Phaselock Technique*, John Wiley, 1966).

At first sight the solution seems easy. A voltage-controlled oscillator generating a square wave can be designed for a wide and linear frequency variation for a variable applied control voltage (see Chapter 11 of EDH). A switching mixer will readily accept a square-wave switching input and the fixed input can be a conventional sine-wave generator. The difference frequency from the mixer may be filtered from the sum frequency by simple low-pass filtering, provided the two oscillators run at much higher frequencies than the required output.

A discriminator of the linearized pump type (see Chapter 13 of EDH) can be used to measure the beat frequency and its output can control the voltage controlled oscillator. This is illustrated in Fig. 6.4.

In this arrangement the VCO is intended to adjust its frequency until the discriminator output voltage is equal to V_{ref}, any departure from this condition resulting in a large control voltage to the VCO. The discriminator characteristic therefore sets the output (beat) frequency for any value of V_{ref}. By adjustment of V_{ref} the output frequency is varied in a well defined manner, oscillator drift being corrected by the loop.

This idea is so tempting as the ultimate solution for wide-range oscillators that there has to be a snag. This usually shows up just when the designer, flushed with apparent success, is demonstrating the range of frequency towards the lower limit. As low frequencies are approached the loop suddenly loses control and the output frequency rapidly rises to a high value and remains there for any setting of V_{ref}.

The let-down is basically due to the fact that a mixer, whether of the switching or multiplying type, delivers among other products an output signal at the difference frequency between its inputs without regard to which input has the higher frequency. In Fig. 6.4 it is assumed that the VCO will always have a higher frequency

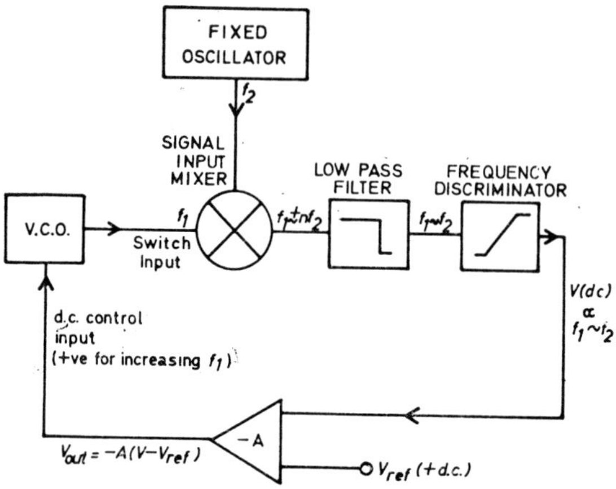

Fig. 6.4 Heterodyne loop

than the fixed oscillator; with the other conditions shown this results in correct direction of drive from the difference amplifier and the loop will operate.

If for any reason the VCO frequency f_1 falls below the fixed frequency f_1 it is clear that the beat frequency now increases when f_1 falls. This means that the 'correction' signal from the difference amplifier drives the VCO in the wrong direction. The discriminator output continues to differ from V_{ref} and the VCO is ultimately driven to its lower limit of frequency. The resulting beat frequency is high, is not related to V_{ref} and loop control has been lost.

What appears to be a simple cure is to limit the capability of the VCO so that it cannot fall below f_1. In practice this is rarely possible; tolerances are such that to ensure f_1 is not reached under any condition would severely limit how near f_1 the VCO could be driven under the intended mode of operation. This therefore restricts the range of beat frequency obtainable, thus degrading seriously one of the most valuable features of the system. A similar comment applies to restriction of the drive voltage to the VCO: because its frequency voltage characteristic is unlikely to be defined to better than a few per cent, a safe limit again cuts down the wanted beat frequency range.

The solution has to be much more subtle if the advantages of a heterodyne system are to be retained. One attempt is shown in Fig. 6.5a; this is by no means fully adequate but is a considerable improvement over the simple methods referred to above.

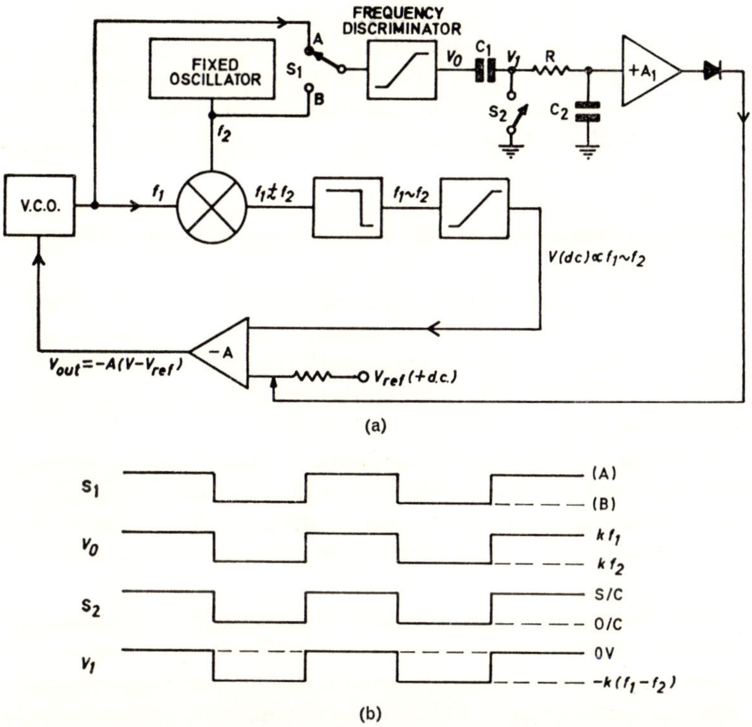

Fig. 6.5 (*a*) Sense loop for Fig. 6.4; (*b*) waveforms

Here, the VCO and the fixed oscillator are coupled alternately to a discriminator by an electronic gate (or relay) which switches at a frequency much lower than that of either signal. If the VCO frequency is higher than that of the fixed oscillator the discriminator output has the form shown in Fig. 6.5b. After synchronous rectification (see Chapter 9 of EDH) and smoothing, a negative voltage is produced. This has no effect on circuit operation, which is already correct, because D_1 is cut off.

If the VCO frequency falls below f_2, the smoothed signal becomes positive and this is coupled by D_1 to the difference amplifier in such a way as to produce a large positive-going control voltage.

Assuming the VCO had fallen below f_1 due to slow drift the correction, having kicked the VCO into the correct region, should now be adequate.

There remain several snags; the accuracy with which the new circuitry can detect when f_1 is only slightly less than f_2 depends ultimately on the electronic switch performance; the adequacy of correction depends on the difference amplifier. This implies that, as in the simple limiting remedy, full exploitation of beat range is not possible. With careful design one can, however, obtain a much better range than by limiting.

Another snag is the odd behaviour of the new loop when its smoothing time is comparable with the beat frequency resulting in erratic oscillation.

There are other more complicated solutions to this original problem. In general each extra complication brings a relatively slight improvement. The designer must watch for these signs and be prepared to call a halt for possible redesign of the system using a different concept.

Remembering the original requirement – wide frequency coverage – it may well be more productive to invent circuit configurations which give directly a wide-frequency oscillator. A possible method is given in Chapter 11 of EDH using a triangle VCO followed by a shaping circuit.

The above system has been dealt with at length to illustrate the problems that arise when a flaw exists in the basic mechanism of a system. Before embarking on detailed design, involving impossible specifications such as VCO limits, the soundness of a scheme should be established by postulating various undesirable yet real conditions: the system must be self-correcting and not rely on avoiding the conditions.

CHAPTER 7

testing

IN theory it would seem unnecessary to build and test new designs conceived and toleranced by trained circuit engineers. Experience has shown, however, that in all but the simplest circuit the designer forgets to consider some feature of the circuit.

This may be due to misunderstanding of the true functioning of the circuit or lack of experience of phenomena that may be virtually impossible to predict by theory.

Often the breadboard tests reveal small deficiencies resulting in modification of some of the circuit values or the addition of protection circuitry. Sometimes a complete redesign may have to be executed because the true system requirements were not met by the circuit although its specification appeared to be satisfied.

The testing stage is extremely important in showing up faults in both principle of operation and in detailed performance, before designs are finalized. Inadequate testing at this time can result in very costly redesigns at a much later date when production failures begin to occur.

The testing of circuits may be undertaken by a technician, an apprentice or a more junior engineer than the designer. It is important, therefore, that the designer know the kind of mistake that is likely to be made by others even if his own test methods are sound. This chapter is devoted to the explanation of errors of this kind which have been reported to, observed by, or perpetrated by the writer. Since it is common practice to leave the design of unstabilized power supplies to the junior or test engineer, some design mistakes and problems in such circuits are also described.

d.c. tests

Whether or not the important signals in a circuit are d.c., the measurement of all direct voltage levels is to be recommended. These levels which should always be recorded are valuable aids for comparison with production units and for servicing.

loading effects

The first mistake, made by almost every beginner, is to forget that the voltage-monitoring instrument may itself cause circuit voltages to change. In the circuit of Fig. 7.1 for example, T_1 collector sits at 8 V. A typical multimeter used on the 10 V range has a resistance of 200 kΩ and its connection between T_1 collector and earth clearly alters this voltage significantly (it falls to about 6.6 V).

Some years ago, this effect became so well known to engineers working on valve equipment that only a beginner would fall into the trap. The multimeters used often had a basic movement of 1 mA (i.e. 1,000 Ω/V f.s.d.) or 0·2 mA (5,000 Ω/V) and circuit loading was considerable. With the more-or-less simultaneous coming of the germanium transistor and 50 μA movements in multimeters (20,000 Ω/V), these loading effects became negligible because resistor values in those circuits rarely exceeded 10 kΩ. With modern planar transistors, however, the situation is reversed and because of their low leakage and high β at low current, resistor values in excess of 1 MΩ are commonplace. The engineer or technician who completed his basic training on germanium circuitry is therefore especially likely to forget this loading effect.

The most striking form of the effect occurs when, in Fig. 7.1, the base emitter and collector potentials are measured separately relative to earth. The impression obtained is that the base sits less positive than the emitter, yet the collector voltage suggests that T_1 is conducting. The reason is that connection of a multimeter to the base (requiring 50 μA or more for f.s.d.) depresses the base potential but the same load on the emitter has little effect since the output impedance there is much lower. If a second meter were used to monitor the emitter potential, still using the first to measure the base potential, a sensible answer would be given (although not the correct, undisturbed potentials).

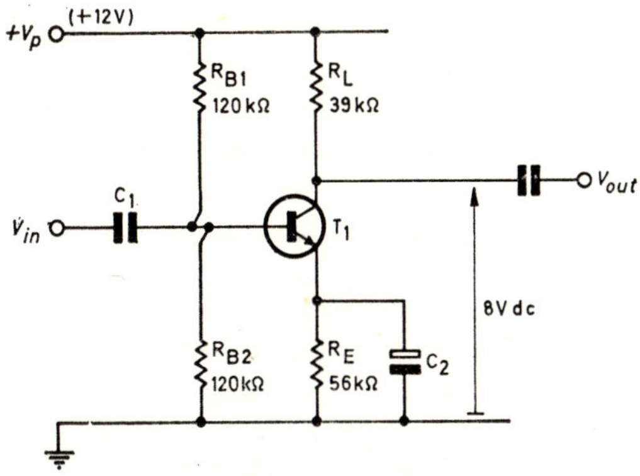

Fig. 7.1 Typical audio circuit

Although the above elementary facts are well known to any engineer the basic point—that attachments may alter circuit conditions—is sometimes missed.

For instance, use is often made of an oscilloscope for these d.c. checks since its input resistance is high. This often means that its 1 MΩ path to earth (fairly typical) is completely forgotten, giving gross errors in low-current circuits. Another more subtle point is that a simple pair of leads several feet long might be used rather than a probe—it seems unimportant in a d.c. measurement. The danger is that hum voltages may be picked up by the lead and although barely noticeable on the screen, they may be sufficient when coupled to the circuit to cause a later stage to overload and upset a feedback path. This can give a completely wrong d.c. level at the monitored point. Another real danger is that the circuit may oscillate when attached to a long lead; the oscillation, like the hum, may be small at the point being monitored but the d.c. levels of the whole circuit may be affected.

The oscillation is usually caused by the capacitance to earth which is introduced by connecting the lead; this causes a change in the frequency response of the circuit which produces instability (see Chapter 8 of EDH).

The introduction of a resistor of about 5 kΩ in series with the lead or probe, placed as near as possible to the circuit being measured, will usually stop this mode of oscillation and still allow d.c. and LF measurements to be made with negligible loss of accuracy. A practical method is to use a probe, clipped to a 4·7 kΩ resistor which has only $\frac{1}{4}''$ leads, and connect the other end of the resistor to the circuit as if it were the probe tip. Note the possibility that the circuit may be oscillating even with no probe connected. If this is so and the oscillating frequency is high, the use of the series resistor, forming as it does a low-pass filter with the capacitance to earth may prevent the oscillation from being seen on the oscilloscope.

Possibly the most difficult loading to anticipate occurs in using a digital voltmeter. These instruments invariably have a very high input resistance (even as high as 20,000 MΩ) and seem to offer the ideal solution to loading problems. Many examples of the instruments, however, present in reality a highly variable input resistance, the mean value of which is very high, but which for short time intervals during the sampling period becomes very low. Moreover, pulse signals are often generated by the digital voltmeter and coupled back into the circuit under test. Naturally, such happenings can alter circuit performance drastically.

The solution to most of the above difficulties is to monitor some significant part of the circuit under test with a separate oscilloscope. Then any disturbance to the correct operation becomes obvious immediately and remedial measures can be taken. Sometimes indirect methods have to be used. In Fig. 7.1 the base potential may be deduced by making current measurements in the base lead and in the R_{b1},R_{b2} chain. The collector voltage can be obtained by measuring collector current.

Current rather than voltage measurements often give more useful information but are tiresome to make, involving the disconnection of soldered joints; if neither connection is 'earthy' there is still the hazard of the injection of ripple voltages.

Another problem is akin to the loading problem voltage measurement; the meter when inserted in series causes a voltage drop. Although the meter movement itself will drop typically only 100 mV a multimeter using the 'universal shunt' principle may drop 0·5 V or more, causing just as much disturbance as in voltage measurement.

When it is known at the design stage that critical d.c. measurements will need to be made, this problem can be eased by adding very small resistors for monitoring purposes. Fig. 7.2 shows the same amplifier somewhat overloaded by monitor points, enabling all currents to be measured by a digital voltmeter with little disturbance to the circuit.

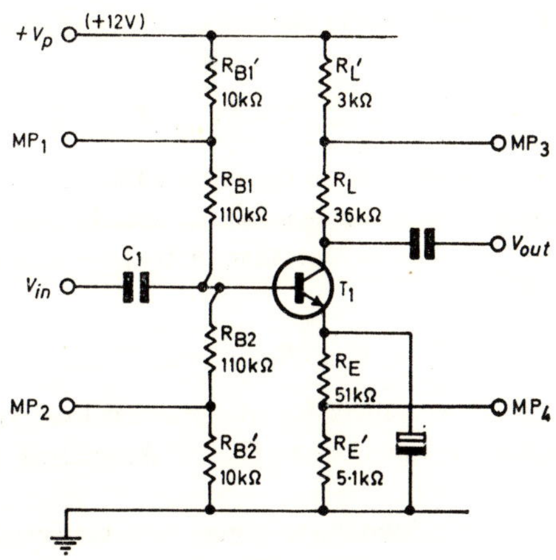

Fig. 7.2 Possible d.c. monitor points

stabilizer problems

One or two pitfalls concerning power supply usage require some knowledge of stabilizer design before their significance is fully appreciated. The output circuit of a typical stabilizer is shown in Fig. 7.3, where T_1 drives the output transistor T_2. In a design situation where the load R_L is outside the power supply designer's control, except for a minimum guaranteed value (here 12 Ω), transistor T_2

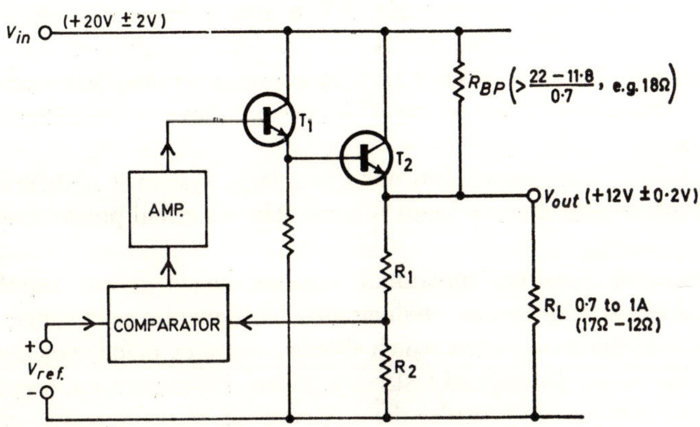

Fig. 7.3 Stabilizer with by-pass resistor

is required to pass any current from $V_{out}/(R_1 + R_2)$, i.e. minimum load current) to

$$[V_{out}/(R_1 + R_2)] + [V_{out}/(R_{2\,min})]$$

i.e. maximum load current. If on the other hand the load is part of a known equipment, it may be possible to specify also a maximum guaranteed value for R_L. In this case, a by-pass resistor R_{BP} may be added such that

$$(V_{in\,max} - V_{out})/R_{BP} < V_{out}/R_{L\,max}$$

This ensures that T_2 conducts for all practical load values and its maximum dissipation is reduced (R_{BP} is dissipating some of the power).

In the given example, load current may vary between 0·7 and 1 A, so R_{BP} may have a value of 18 Ω. When V_{in} is at +22 V, R_{BP} passes 0·56 A and T_2 emitter current varies between 0·14 and 0·44 A over the whole range of load currents. This reduces T_2 dissipation from 10 to 4·4 W at maximum load and is therefore a most useful modification.

Returning to measurement errors, it is evident that a stabilizer designed in this way as part of a large system will give an excessive, unstabilized output, full of ripple, if used on a load current which is less than the normal by-pass current $(V_{in} - V_{out})/R_{BP}$.

Again referring to Fig. 7.3, even if no by-pass resistor R_{BP} is used, an external load connected between this output and a higher voltage will produce the same result. This often arises in large systems where a circuit requiring, for example, 8 V is strung between, say, +12 and +20 V supplies. No doubt the designer has allowed for this 'backward' current into the +12 V by ensuring it is less than normal loads between +12 V and earth: loads may have to be added for testing.

In testing only one or two units of a large system it is therefore advisable to add dummy loads if the system's normal power supply is to be used.

Whenever possible, individual variable bench power supplies should be used for testing small circuits. This enables each supply to be varied separately, a test which should always be made even if not specified in any formalized test requirement. Having set up nominal supply voltages and made the appropriate measurements, each supply should be varied by about ±10 per cent, and any gross performance changes noted. Most circuits will then have a different performance, possibly outside the normal specification if supply lines in the actual equipment were guaranteed better than ± 10 per cent. Complete failure to operate, however, is a danger sign that the design is marginal in some respect and should be re-checked. Usually an operating current is either too low, resulting in cut-off under extreme supply conditions, or too high, causing bottoming. Some designs involving closely defined trigger levels may have every right to fail if supply lines vary by this amount; but the designer would be aware of this and could safely ignore this result.

It is in many ways a good idea, when a circuit uses several power supplies, to add a master switch between the circuit and the supplies. Psychologically it makes it much more likely that the test engineer will take the trouble to switch off before applying a soldering iron which may be either earthed, or connected by several hundred pico-farads to the mains supply (if its earth lead has been removed). It also ensures that, after a modification is made and power re-applied, conditions are just as before – no supply line now present which was accidentally absent before – no variable control accidentally touched in switching on or off.

In other ways this master switch seems to give the wrong impression that the lines are being connected and disconnected simultaneously. This is very unlikely to be true for a manual switch in terms of the few milliseconds needed to destroy a semiconductor; a circuit that cannot accept without damage some supplies in the absence of others is therefore quite unsafe. Conversely, if the switch is so accurate that simultaneity *is* achieved, then in the circuit's final environment the same conditions are unlikely to obtain. A design that cannot accept separate supply switch-on could therefore be passed as satisfactory and not be revealed in its true light until final commissioning of an entire system is carried out.

The safe method is to use the master switch, but still check the circuit for its ability to withstand individual supply switch-on in all combinations while leaving the master switch closed.

Another trap that occurs in using modern bench supplies is concerned with the supplies' overload protection. Most are now of the current-limiting type whereby an excess current from the supply causes it to cease normal stabilization and become a constant-current source until the overload is removed. Earlier supplies and many special-purpose supplies use a non-reversible trip mechanism so that an overload must be removed and then the supply reset manually. To indicate the presence of an overload in the current- limiting type of supply a neon indicator is usually provided.

The danger, which can occur in Class AB, B or C amplifiers or in power pulse circuits is that an overload during only part of a cycle operates the lamp for too brief a period to be readily visible. Since on each overload the power supply becomes very high impedance, circuit performance is likely to be affected greatly.

In dealing with circuits taking power in pulses it is therefore good practice to monitor, by means of an oscilloscope, the supply lines. If overloading takes place a large fall in supply voltage will appear at the instant of overload.

When a circuit is powered by its own rectifier supply from the mains, it is always necessary to check its behaviour with specified mains variations, sometimes ± 5 per cent, often ± 15 per cent if mains tap adjustment is not allowed. The obvious method is to connect a continuously variable voltage-transformer between the mains and the circuit.

The pitfall here is the assumption that the supply reaching the circuit rectifiers is of the same low impedance as the mains. In low-powered circuits this does not cause errors but when the rectified load is in excess of 100 W the pulses of rectifier current taken at the peaks of each half cycle cause large drops in the windings of the variable transformer. Clearly, the resulting loss of output occurs only at the peaks and the usual average-reading multimeter-check on the alternating voltage leaving the transformer shows only a slight loss. The engineer merely turns the variable control to a higher setting until the intended voltage appears on the multimeter. The average reading is now back to normal but the peak value is much too low and the circuit rectified voltage (if a peak rectifier) is also low. The simple check is to monitor the waveform from the variable transformer when on the actual load presented by the circuit. If the sine wave is appreciably flat-topped, a variable transformer of heavier rating should be used. If this is not available, the primary winding of an adequate fixed but tapped transformer may be used and the input or output tapped down the winding, again monitoring the output waveform. In practice *a normal transformer is much lower in effective resistance than a variable type of similar rating.*

dynamic tests for stabilizers

In the design of high-gain stabilizer loops one of the most difficult problems is to ensure freedom from instability round the feedback loop. The principles are well known but when various values of load current and all transistor β, f_T and C_{cb} tolerances are taken into account, determination of the loop transfer function becomes virtually impossible unless an extremely large dominant time constant is used (see Chapter 8 of EDH).

It is therefore extremely important in testing the prototype that complete loop stability is checked. It is not sufficient merely to look for actual oscillation at various load currents, because the circuit may be very close to oscillating. The circuit when produced in quantity may well (and very often does) turn out to be oscillatory in a large proportion of cases. The classic test for this purpose is a 'gulp test' in which the load current is rapidly switched, electronically, from minimum to maximum while observing the stabilizer output waveform. This is a most searching test and any stabilizer that is even remotely near oscillation produces spectacular rings at each switching of the load. The waveform is also useful in stable circuits in showing the recovery time and output resistance after recovery.

Figure 7.4a shows the circuit of a suitable gulp tester for currents up to a few amperes. For practical use it is best powered by its own battery since it is then not tied to a common earth system and can be used 'floating' on any supply, positive or negative with respect to earth. The typical waveforms in Fig. 7.4b show a stable and an oscillatory stabilizer when subjected to this test. The p.r.f. is not critical provided the half-period exceeds the settling time, and a stabilizer showing no ringing after the inevitable initial transient can be declared safe with confidence.

d.c. conditions in logic circuits

Brief mention must be made of a phenomenon that has been reported so often to the writer that it must rate as the most common blunder of all.

The story is always the same: a system containing logic has failed to operate and a check has therefore been made on the d.c. conditions. On checking any of the binary circuits it is found that both transistors

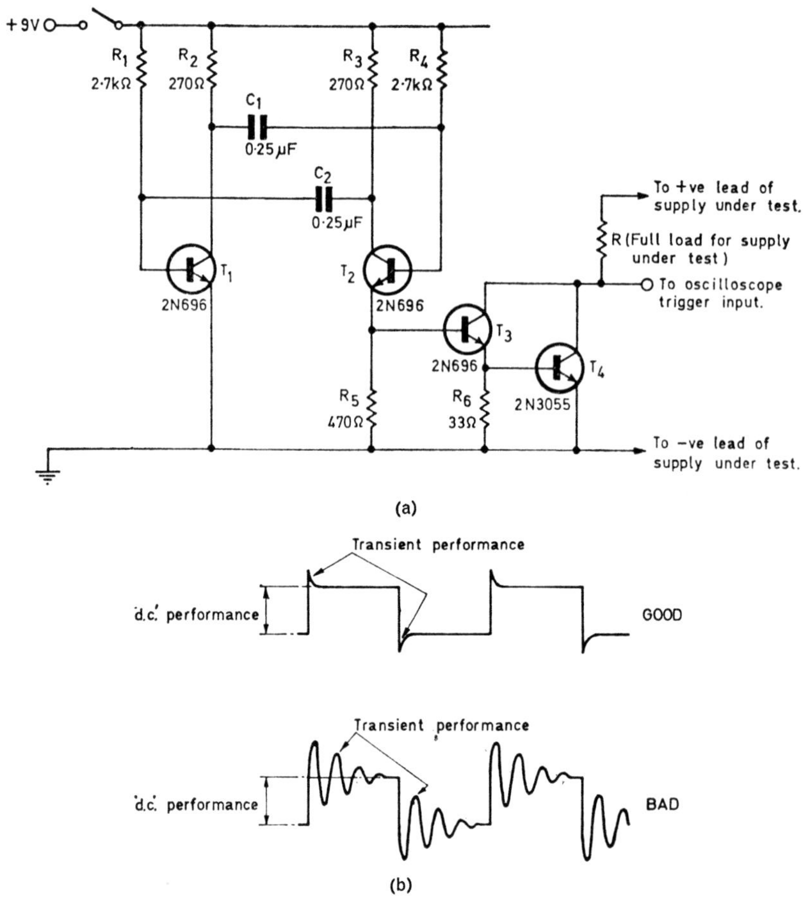

Fig. 7.4 (*a*) Gulp tester for power supplies; (*b*) voltage waveforms across terminals of supply under test

in every binary are bottomed. Replacement of the transistors (and eventually every circuit element!) will not effect a cure.

What happens is that the collector voltages of the two transistors are measured *one-by-one*, using a multimeter or oscilloscope which under quiescent conditions presents a safe load. Unfortunately, the mere capacity of the test instrument is sufficient to trigger the binary if the collector being examined is of high impedance, i.e. cut off. If, on the other hand, the collector is already bottomed, no triggering occurs. Each collector therefore seems to be 'on' when measured and for some reason the impression felt is that the measurements are simultaneous. The reason for system failure is invariably another unrelated effect. To check whether a standard two-transistor binary is in reality a two-state circuit, two simple methods are available.

A multimeter or oscilloscope can be connected to one collector, normally triggering this transistor 'on' giving a low voltage. Leaving this connection undisturbed the binary can now be made to change state by short-circuiting the base-emitter of the conducting transistor, by short-circuiting the collector-emitter of the cut-off transistor, or often by holding the blade of a screwdriver and touching the collector of the cut-off transistor. The whole process can then be repeated with the meter connected to the opposite collector so that the 'on' and 'off' voltages of each transistor are measured.

Alternatively two meters may be used, one on each collector, and the binary triggered by momentarily disconnecting and then re-connecting one of them.

Naturally, a binary with added buffer stages, e.g. 930 series DTL, JK flip-flop is immune from the problems just described.

power measurement in switching circuits

Mistakes are frequently made in power measurement in switching circuits. Where sinusoidal waveforms are involved the engineer is immediately aware of the possibilities of phase angle, and makes a clear distinction between real and reactive power. When mean levels are not zero and can therefore be read on d.c. instruments this caution tends to disappear and the assumption is made that mean power is the product of mean volts and mean amperes.

Figure 7.5 illustrates convincingly the effect in question. Switch

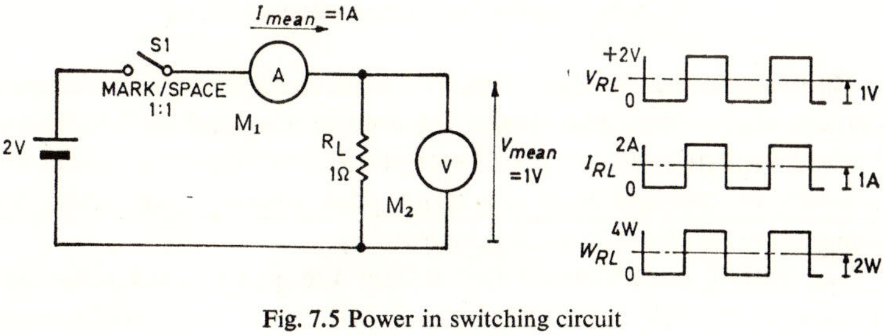

Fig. 7.5 Power in switching circuit

SW_1 is oscillating continuously with a mark/space of 1/1; meter M_1 reads direct current and M_2 reads direct voltage. Since for half a period R_L has 2 V applied to it and 2 A then flow through it, the mean voltage read on M_2 is 1 V and the mean current on M_1 is 1 A.

It would now be a common assumption that the power dissipated in R_L is 1 V × 1 A = 1 W. Closer examination shows that during the 'on' time of SW_1, power in R_L is 4 W; this gives an average of 2 W – the correct figure.

The rule in calculating power in such circuits is that the average may be taken only of the power waveform, unless either the current or voltage is constant. Battery power may be calculated by noting that the voltage is constant at 2 V, and the mean current is 1 A, giving 2 W. Note that, since no power can be lost, this must equal the power in R_L.

Since circuit testing often requires the construction of a suitable unstabilized power supply, it is worth emphasizing some of the problems associated with rectifier systems such as Fig. 7.6. Lack of attention to ratings here may result in the intended tests being overshadowed by violent explosions. The operation of this type of circuit is simple in principle but very complicated to calculate accurately, *C* losing charge continuously into the load and being recharged during a fraction of the period. A complete analysis is given in Chapter 1 of EDH and the values given in Fig. 7.6 are typical for a TV receiver power supply.

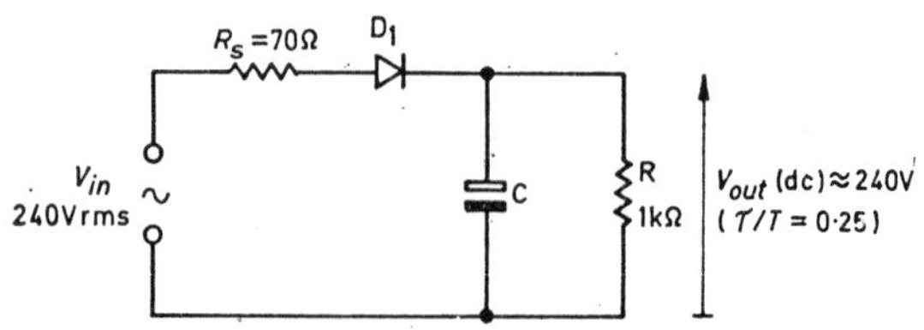

Fig. 7.6 Peak rectifier, giving high dissipation in R_S

R_S protects the rectifier from excessive switch-on and recharging current surges and also limits the output on load to 240 V d.c. Failure to include R_S results in rectifier failure as soon as switch-on happens to coincide with the input peak (usually only when the designer switches on at a demonstration).

The pitfall occurs when calculating the power rating for R_S; a common procedure is to work out I^2R_S ,where I is the load current

(240 mA), giving 4 W. In fact the current waveform consists of pulses having a mean value of 240 mA but with a duty cycle of 1 : 4. The voltage waveform across R_S is similar in form and a correct power calculation (remembering that these waveforms are not rectangular) gives a peak power of about 120 W and mean power of 30 W.

The discrepancy in these power ratings between a casual calculation and a correct one is surprisingly large and even an apparently generous factor of safety could still give very short resistor life. This example is in no way exaggerated and the figures given are typical.

In selecting a reservoir capacitor for such peak rectified power supply circuits it seems a simple matter to decide on the required capacity to give the necessary low ripple and choose a d.c. working voltage with some safety margin.

An equally necessary but often neglected rating is the ripple current capability of the capacitor. The alternating ripple current causes temperature rise by dissipating power in the effective series resistance of the capacitor; if this temperature rise is excessive the contents of the capacitor case are deposited over the rest of the equipment in a noisy and spectacular manner.

There is one excess ripple current condition that can only be protected against by a fuse or circuit breaker; this happens when one or more of the rectifiers become short circuit thus applying the full a.c. input to the capacitor. This usually comes about because R_S has been omitted, the diode being destroyed by excess forward current at switch-on or on each recharge pulse.

During normal operation the ripple current is approximately 2·2 × d.c. load current, almost regardless of other circuit conditions. This is readily proved by reference to Fig. 7.7 in which the recharge period is assumed small compared with a period of ripple.

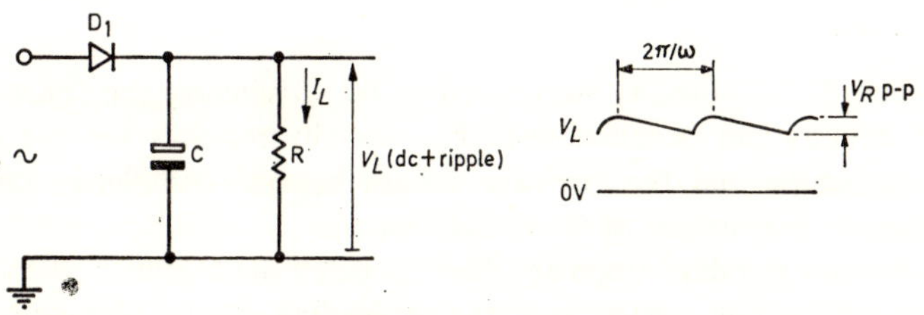

Fig. 7.7 Ripple in peak rectifier

Assuming that V_L is large compared with R_V pk-pk, the load current I_L is constant at all times. This discharges C at a rate given by $dV/dt = I_L/C$ so that

$$V_R \text{ pk-pk} = \frac{2\pi}{\omega} I_L/C$$

where ω is the angular ripple frequency. If the further (pessimistic) assumption is made that the ripple waveform is sinusoidal, then

$$I_C = \omega C V_R \quad \text{and} \quad I_C(\text{r.m.s.}) = 2{\cdot}2 I_L$$

In the example of Fig. 7.6 the ripple rating of C must therefore be about $2{\cdot}2 \times 240 \times 10^{-3}$, i.e. about 0·5 A.

Manufacturers of electrolytic capacitors quote ripple ratings for those types intended for reservoir use; if no figure is quoted it is not safe to assume one.

There is an interesting paradox with regard to reservoir capacitor selection. If space is lacking and the output ripple from the supply is unimportant, it appears sensible to use a small-valued reservoir. When its ripple rating is calculated, however, this is found to be, as usual, about twice the direct load current. Examination of available capacitors then shows that to obtain a small-valued capacitor of that ripple rating, a larger voltage type has to be used than would be needed purely from d.c. voltage rating considerations. This capacitor turns out to be just as bulky as the obvious choice having just enough working voltage, large capacitance and the same ripple rating. Since the latter gives a bonus of lower ripple it may as well be used. The inescapable fact is that size is tied to ripple rating, and the only clear way to take advantage of an easy ripple specification is to use a series choke rather than a peak rectifier system, the choke being no larger than the ripple specification demands.

a.c. tests

When d.c. conditions are proved to be satisfactory the functioning of circuits can be examined. This usually requires the use of an oscilloscope and the engineer should become completely familiar with the adjustment of these instruments.

Several standard traps are likely to catch the beginner when using an oscilloscope, apart from the misleading results obtained when an uncalibrated control is not set to its one calibrated position.

When using a probe, which is mechanically convenient and is connected to the oscilloscope by a screened cable, it is not permissible to insert an extra length of cable. The appearance of the viewed waveform and the over-all frequency response will be greatly affected. Similarly, shortening the cable, which may seem reasonable if it is broken by excessive bending, also affects response since the inner conductor is *not* a normal low-resistance wire.

Means are usually provided for adjusting probe response by a calibrating square-wave source which is part of the oscilloscope, and a trimmer in the probe. When examining waveshapes it is essential that the probe is set up on the actual *Y* input voltage range which is to be used.

When a twin-beam display is used to examine two related waveforms simultaneously it is important to ensure that each beam is triggered in the same manner. This is automatically done when a chopped beam system is used, in which the trace jumps between the two signals at a rate that is much faster than the scan rate. At high time-base speeds, the chopping rate required to avoid a dotted-line appearance would be too difficult to achieve; alternate sweeps are then made for each signal. The danger is that if the time base is internally triggered, each signal initiates its own sweep; a waveform on the second *Y* channel which is identical to that on the first but displaced in time therefore appears directly below the first waveform.

When the waveforms are not identical in shape, the turning of the time-base trigger level control usually causes the apparent time relationship of the beams to change and almost any relative positions can be obtained.

Since the whole object of a two-beam display is to examine time relationships between two signals, this triggering mode is completely unsatisfactory. The cure is to use the 'external-trigger' position on the trigger selector switch and to connect a suitable signal (which may be one of the two being examined) to the trigger input terminal.

This brings up another pitfall: trigger input circuits usually present a much heavier load than the *Y* inputs. Often an attenuating probe can be used (or a small series capacitor if a pulse rather than a slow trigger signal is used) which greatly reduces the loading. In other cases the trigger waveform can be obtained from a part of the circuit under test that is capable of accepting the load. Failure to observe these precautions can result in apparently poor performances from the circuit.

Whenever signal and d.c. levels are suitable the oscilloscope should be d.c. connected. The voltage relationships between circuit waveforms usually give the clue to any unsatisfactory performance. The margin by which a transistor avoids cut-off or bottoming during part of a cycle can often be seen by comparing collector and emitter or collector and supply-line voltages. It must be remembered, as remarked earlier, that a directly coupled oscilloscope presents a leakage path to ground (usually 1 MΩ on the oscilloscope, or 10 MΩ if a 10:1 probe is used.)

When accurate measurements have to be made, electronic voltmeters are often used. Although invaluable for certain sine-wave measurements especially in the 50 to 1,500 MHz band, their mode of operation limits their usefulness considerably. Some instruments measure the average rectified value; others measure the peak or peak-to-peak value; a few measure 'true r.m.s.'; all are normally calibrated in terms of r.m.s. value.

This means that only sinusoidal waveforms will be measured correctly (except when using a 'true r.m.s.' instrument). The mistake sometimes made is that the output of an amplifier circuit handling sine-waves is measured by an electronic voltmeter without monitoring the waveform on an oscilloscope. It is quite common in such amplifiers for distortion to occur only at higher frequencies, owing to capacity loading, at very low levels due to cross-over distortion, or only at high levels owing to overloading. It is advisable to check the waveform continuously so as to be sure of the validity of the voltage measurement and to confirm that the voltmeter itself causes no degradation in performance.

temperature tests

Whether or not a circuit has to operate in a high or low ambient temperature, it is good practice to perform temperature tests to show up any marginal points in the design. This is usually carried out in the laboratory by using a small thermostatically controlled oven.

When slight performance variations are being measured, rather than simple monitoring to check for complete failure, some precautions must be taken to obtain realistic results.

The equipment under test must be in the oven long enough to reach the test temperature; this takes much longer than reason would suggest. A circuit raised from normal laboratory temperature to 50°C should be left for about an hour before stable operation can be expected. If the test circuit is coated or potted this time must be increased to several hours.

When the correct performance of the circuit at different temperatures depends on matching of component temperature coefficients (such as V_{be} of transistors, resistor chains, inductance coefficients which match capacitance coefficients) then the components must be mounted in such a way as to reach the same temperature. This is an obvious requirement but is not easy to achieve because air movement causes a temperature differential if the components are not enclosed in the same container.

If no special precautions are taken, a temperature differential of a few degrees per inch can occur, and this can completely upset a critical circuit.

A complete cure to this problem is achieved by enclosing the relevant components in a plastic or metal box which simply stops air movement between them. The box must naturally be used also in the final equipment which will experience just as many air movements as provided by the test oven. As a by-product, the box also acts as a shield for radiated heat which may be emitted by a nearby circuit component and received unequally by the matched components.

If the tests involve the accurate measurement of low voltages then great care must be taken to avoid the formation of a sensitive spurious thermocouple. This happens if the measuring leads contain two junctions at different temperatures between dissimilar metals and can be confirmed by measuring a voltage (such as across part of an earth track) known to be zero. The cure may consist in changing the materials on one side of the junction or in repositioning junctions until the above check proves satisfactory.

Another problem is the measurement of the temperature at which the test is made. The usual suspended thermometer is not accurate for two reasons. Firstly its reading depends on heat which may be radiated from circuit elements (if placed near the circuit). Secondly, even if there is no radiated heat, it reads the temperature at some point in the oven which may be a hot or cool spot or may be quite different from the temperature experienced by the circuit due to air movement. There is no real solution to this problem but for critical applications it is evident that several degrees margin must be allowed when testing to be confident of successful operation in the actual environment.

One pitfall worth mentioning here has caught unawares many

experienced engineers. A digital circuit containing a counter chain which counts for say, 1 min, is found on an oven test to fail at exactly the maximum temperature it is required to withstand. The usual reaction is of deep concern as to whether this makes the design just acceptable or not.

The truth is that in performing the test, the oven is naturally set to the maximum temperature required. Just as this is reached the oven thermostat operates and the resulting interfering pulse triggers several stages of the counter giving a completely wrong count. Having deduced the cause, the engineer of low cunning can of course predict to his assistant that if tested 15°C higher, the circuit will then work 15°C higher and again just fail! Whether such response to transients will be important in actual usage of the circuit depends on the nature of the system. If a timing sequence is involved this susceptibility could be disastrous; if on the other hand a continually updated measurement is being made, as in a bench frequency-counter, the occasional wrong answer could be insignificant.

CHAPTER 8

examples of faulty design

THIS chapter describes design errors that the more experienced designer may commit during an 'off' day. They are not merely mistakes of arithmetic but are cases where a momentary lapse of reasoning can lead to disaster even though circuit components have apparently been given logical values.

In these examples the calculation of circuit values will not be explained except where directly relevant to the errors under discussion, and those values given can be taken as typical.

power amplifier

The type of circuit of Fig. 8.1 is now almost a standard arrangement for power amplification. Its properties have been described in detail

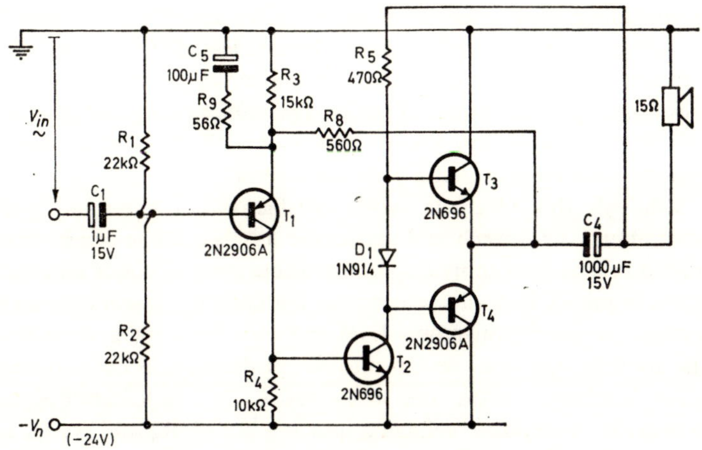

Fig. 8.1 Audio amplifier

from the audio-engineering viewpoint by several authors, notably by Dinsdale in his excellent *Wireless World* series.

One of its snags is that a short circuit across the load – usually a loudspeaker – allows enormous currents to flow in T_3 and T_4 if a large input signal is present, leading to the failure of at least one of these transistors. Since a short-circuited load is a real possibility, as many hi-fi enthusiasts would sadly agree, it seems reasonable to limit the available current by adding R_6 and R_7 as shown in Fig. 8.2.

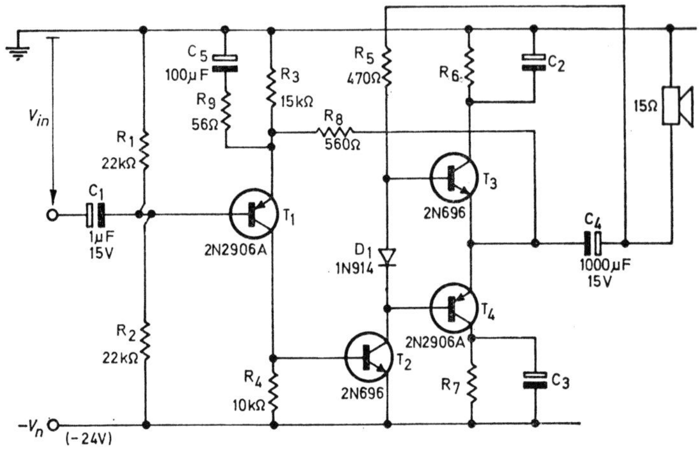

Fig. 8.2 Protection which restricts output capability

By including these resistors in the collector circuits of T_3 and T_4 the normal performance appears to be unaffected; C_2 and C_3 must naturally be added to decouple the collectors so that bottoming does not take place on signal peaks. The values of R_6 and R_7 could each be 15 Ω since it is known that T_3 and T_4 survive in normal operation the current that a 15 Ω load takes.

Although this idea can work satisfactorily for normal audio programmes, i.e. speech and music, the mistake which can be made only too easily is to apply it to a case where normal power amplification is required. In audio-programme material, the signal only rarely reaches the maximum power level and then only for a short time. The amplifier has to be designed to accept these peaks but is running at low level most of the time. If C_2 and C_3 in Fig. 8.2 are designed to prevent appreciable voltage appearing across R_6 and R_7 in, say 10 msec, but allow T_3,T_4 to bottom in the event of a combination of high input level and short circuit load within, say 30 msec, damage will usually be prevented and most programme material will be unaffected. If the circuit is now used as a power amplifier which operates continuously near maximum output, the output swing will be drastically limited.

This is one of the cases where unequal charge and discharge paths exist, causing mean levels to change to achieve equilibrium (see Chapter 1). C_2 receives charge on each positive half cycle from T_3 collector current, but loses charge only through R_6 giving a standing voltage on T_3 collector well below the quiescent no-signal level. A larger value of C_2 merely prolongs the time taken in establishing equilibrium. The actual level for a large C_2R_6 product is easy to calculate, since for a peak output signal $\hat{V}_{out}$, T_3 collector current consists of half sine waves of peak amplitude $\hat{V}_{out}/R_L$; this flows through R_6 and the waveform is smoothed by C_6 giving a mean drop across R_6 of $V_{out}R_6/\pi R_L$. A similar expression holds for R_7.

Limiting occurs when T_3 bottoms on each peak and this occurs when

$$\hat{V}_{out} - \tfrac{1}{2}V_n = -\hat{V}_{out}R_6/\pi R_L$$

i.e. when

$$\hat{V}_{out} = V_n/\left[2\left(1 + \frac{R_6}{\pi R_L}\right)\right]$$

With the present values $\hat{V}_{out} \approx 9$ V; without R_6 and R_7, $\hat{V}_{out} \approx 11$ V. Under conditions where bottoming occurs on every peak the charge equation becomes invalid since, during the bottoming interval, C_2 is driven in the opposite direction to the normal flow of T_3 collector current.

Attempts to avoid this charge accumulation and yet retain the protection are doomed to failure; for example, an extra capacitor linking T_3 and T_4 collectors merely lengthens the smoothing time constants.

Summarizing, C_2 and C_3 can only prevent transient voltage drops appearing across R_6 and R_7; whatever happens, the standing voltage on these resistors must correspond to the mean collector currents of T_3 and T_4 which depend on the rectified mean load current. This may be very small in a typical audio signal even though peaks are high. The form of protection under discussion may therefore be quite satisfactory in such cases but not in continuously high level applications.

A similar situation arises when the other main snag of this circuit is encountered—cross-over distortion. In Fig. 8.1, T_3 and T_4 are biased off in the quiescent condition, since the drop across D_1 is less than the minimum value of $(V_{be3} + V_{eb4})$. Sometimes an adjustable resistor is added in series with D_1 and its value set to turn on T_3 and T_4 to the extent of a few milliamperes. The choice between the two evils is most difficult. Without a resistor, T_3 and T_4 remain cut off near the zero level of signal causing cross-over distortion; with the resistor, the current in T_3 and T_4 is badly defined and increases as the temperature rises (because the two V_{be}'s are opposed by a single diode drop). There are practical difficulties with the resistor also: it must have a maximum value which is sufficient for any combination of T_3, T_4 and D_1. If on switching on a newly built circuit the resistor happens to be set at maximum and the V_{be} drops require only the minimum value, T_3,T_4 may be turned on hard, causing their destruction.

In view of this, the attempt is often made to define a small quiescent current in T_3,T_4 by inserting emitter resistors R_{e3} and R_{e4} (Fig. 8.3). This enables the resistor R_6, in series with one or even two diodes, to set the current with reasonable accuracy, which implies ± 100 per cent in this type of circuit (it is of little importance whether the standing current is 2·5 or 10 mA). The resistors may be split as shown to obtain the correct d.c. level for the d.c. feedback loop.

It turns out in most practical examples that R_{e3} and R_{e4} are too

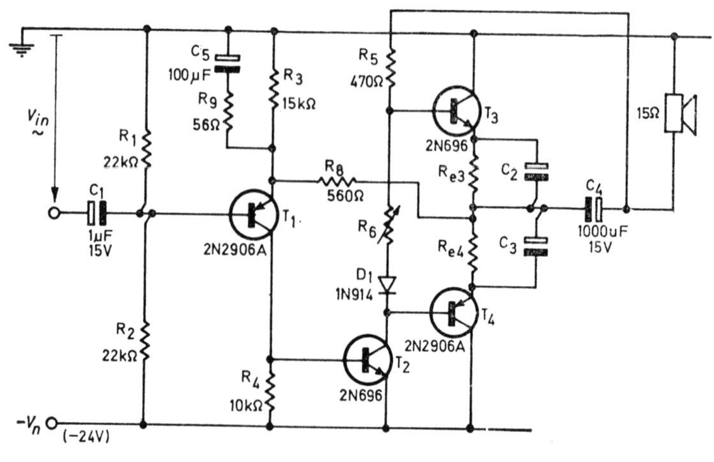

Fig. 8.3 Unsuccessful attempt to avoid crossover distortion

large to feed the output load. Coupling capacitors C_2 and C_3 are therefore added so that the output impedance is reduced.

This appears at first to be a convenient solution but on examining the charging conditions of C_2 and C_3 it becomes clear that this is another case in which a.c. levels depend on signal amplitude. With large input signals, C_2 will be charged heavily by T_3 emitter current on positive signal half cycles but can only discharge through R_{e3} on negative half cycles since T_3 is almost at cut off. The inevitable result is that the junction of C_2 and T_3 emitter becomes more positive until equilibrium is reached. From the point of view of the load, only extreme positive and negative tips of the signal are received if R_{e3} and R_{e4} are large; with normal values more of each half cycle is passed but there is severe cross-over distortion (largely concealed by the negative feedback).

The addition of a capacitor directly between the emitters gives no improvement, and the whole concept of defining the current has to be valued on the assumption that C_2 and C_3 do not help. This simply implies that R_{e3} and R_{e4} must be low enough not to require C_2 and C_3, usually giving inadequate definition of T_3,T_4 current.

pulse/ramp generator

The next example illustrates a common mistake in which the designer's view of the mode of operation of his circuit is at fault.

The idea (see Fig. 8.4) is to generate a sawtooth waveform of defined amplitude and period (waveform *a*) and simultaneously produce a pulse waveform as shown in waveform *b*. This is a useful

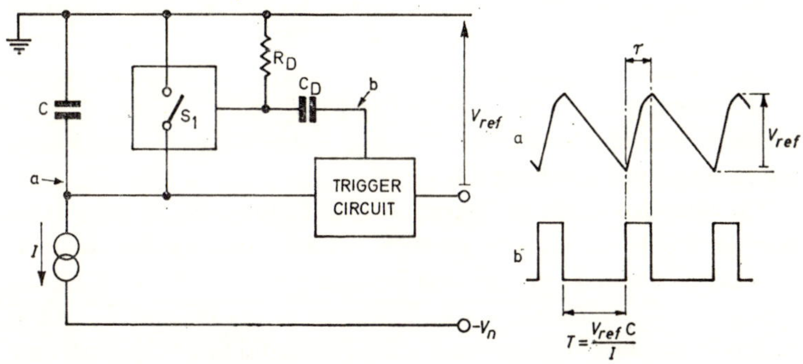

Fig. 8.4 Pulse/ramp generator and intended waveforms

combination when driving CRT deflection amplifiers with the saw-tooth and blanking circuits with the pulse. The method shown is satisfactory in principle and at first sight easy to execute. Capacitor C charges linearly from the constant current source until its voltage reaches the trigger level of the trigger circuit which is equal to V_{ref}. The resulting output excursion is differentiated and operates electronic switch S_1 which discharges C to zero. S_1 now opens either because the trigger circuit has changed state again or because the differentiating time governed by $C_D R_D$ has expired. C now charges again and another cycle begins.

Figure 8.5 shows an apparently satisfactory circuit in which ZD_1, T_1 and R_2 act as the constant current source, T_3, T_4 and T_5

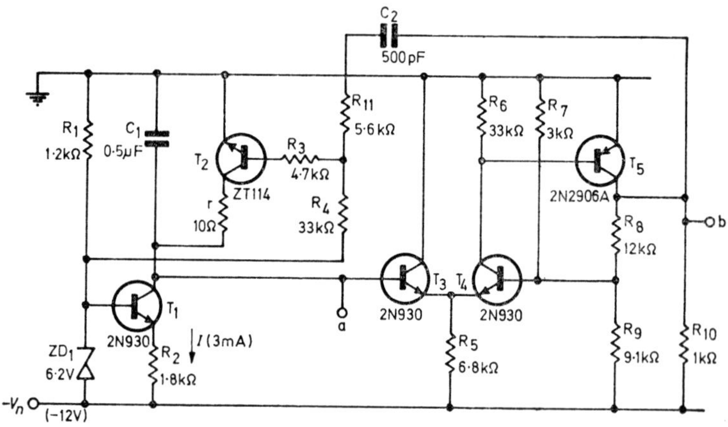

Fig. 8.5 Practical form of circuit of Fig. 8.4

form a trigger circuit (see Chapter 9), C_2, R_3 and R_4 form the differentiator and T_2 is the transistor switch. The values are calculated to give a lower trigger level of about $-4{\cdot}5$ V and the ramp time would be expected to be given by $t = VC/I$ where $V = 4{\cdot}5$, $C = 0{\cdot}5 \times 10^{-6}$, $I = (6{\cdot}2 - 0{\cdot}7)10^{-3}/1{\cdot}8 = 3 \times 10^{-3}$, i.e. $t = 0{\cdot}75$ msec, the ramp amplitude being 4·5 V.

The intended mode of operation is as follows. C_1 charges at the rate of I/C V/sec until it reaches $-4{\cdot}5$ V. During this time T_3 conducts, T_4 and T_5 are off, and T_4 base rests at $-4{\cdot}5$ V. When the voltage on C_1 reaches $-4{\cdot}5$ V, T_3 begins to cut off and T_4 and T_5 begins to conduct. This action is cumulative and in a very short time, limited mainly by stray capacitance and transistor f_T, T_5 becomes bottomed. T_4 base potential becomes about $-2{\cdot}5$ V and T_3 is now cut off by a margin of 2 V. The resulting voltage step at T_5 collector is differentiated by C_2, R_3 and R_4, and T_2 is bottomed, discharging C_1. On the way to zero, V_{C_1} passes through $-2{\cdot}5$ V so that when finally discharged the trigger circuit is back to its original state with T_3 conducting and T_4 and T_5 cut off.

This explanation may sound plausible but contains a basic flaw which results in wrong timing and amplitude. When T_2 is turned on due to the bottoming of T_5 it is true that C_1 discharges towards zero. However, *as soon as* V_C *reaches* $-2{\cdot}5$ V, T_3 *is turned on* and T_4 and T_5 are cut off. When this occurs T_2 is immediately driven to cut off by C_2, and C_1 starts to charge negatively once more. The excursion on C_1 is therefore 2 V, equal to the backlash of the trigger circuit. There is a tendency to feel that, because T_2 is a powerfully driven switch, C_1 will discharge to zero before anything else can occur. This is only so if the trigger circuit is incapable of changing state in the time required to complete the discharge. In any practical circuit the trigger circuit will be faster unless special measures are taken.

Because the excursion of V_c is only 2 V the timing is wrong in the ratio 2 : 4·5 and is 0·33 msec instead of the expected 0·75 msec.

There are several cures that suggest themselves once the problem is made clear.

One approach is to accept the actual mode of operation as satisfactory and design the trigger circuit to meet the original amplitude and time requirements. In this case the trigger levels would have to differ by 4·5 V instead of 2 V and this can be achieved by changing R_7 and R_8 (see Chapter 9 for method of calculation). It is not essential to use a differentiating circuit and direct resistive coupling from T_5 to T_2 is possible. A remaining defect then is that C_1 does not discharge to zero, but only to the upper trigger level. Although it would seem possible to make this level quite close to zero by using a very large value of R_9 this would be bad design practice: unbalance in the V_{be} of T_3 and T_4 could mean that T_3 never turns on even with V_{C_1} equal to zero.

The effect of failure to trigger back depends on whether C_2 is used. If it is, then T_2 turns off, C_1 charges until T_1 bottoms because the trigger circuit is already in the 'T_3 off' state. If C_2 is not used, then T_2 remains bottomed and C_1 remains close to zero. In either case the circuit fails.

The method by which the amplitude is determined by backlash is therefore not suitable if a ramp is needed to start accurately from zero.

An alternative is to slow down the speed of the waveform coupled to the trigger circuit so that C_1 discharges to zero before the trigger circuit is aware of it. It is of course futile to slow down the action of the trigger circuit itself by, for example, adding a capacitor across R_6. This would also slow down the output step on T_5 collector and result in slow discharge of C_1, defeating the object. A suitable method is to add a resistor in series with T_3 base and a capacitor from T_3 base to earth, e.g. 4·7 kΩ, 0·02 μF. Provided T_2 passes enough current to discharge C_1 to zero before the voltage on T_3 base reaches −2·5 from −4·5, full amplitude and correct timing is obtained. One snag is that the charging curve for C_1 is now slightly modified and the flyback time of the ramp is dependent upon the new components.

Another method is to insert a series diode D_1 with resistor R_x and shunt capacitor C_x to the negative line in the T_5/C_2 connection (see Fig. 8.6). This peak rectifying circuit 'holds' the effect of T_5 bottoming long enough to discharge C_1, but the discharge time constant R_xC_x is short enough to allow C_x to reach −12 V before the next trigger occurs, i.e. $4C_xR_x < 0{\cdot}75$ msec. It suffers from the snag, however, that there now tends to be a waiting period between ramps.

All these suggestions are made to illustrate the principle; none of them can be regarded as a good solution to the original problem

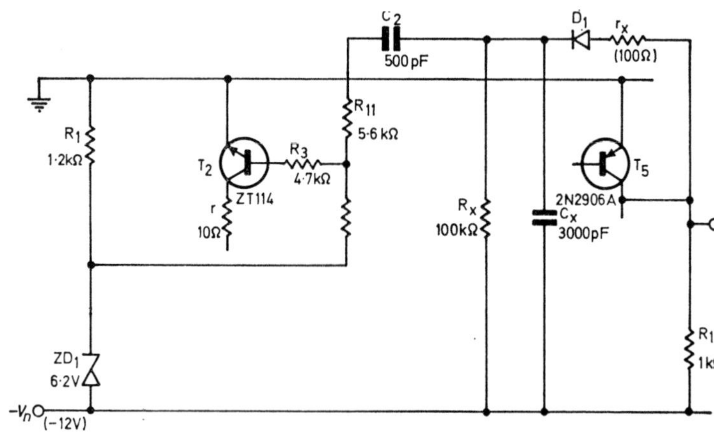

Fig. 8.6 Modified form of circuit of Fig. 8.5

and they are generally unsatisfactory if variable frequency operation is required. The only wholly acceptable course of action is to conceive a different system which avoids the problem completely.

d.c. stabilizer with faulty protection

When a voltage stabilizer circuit is designed it is usually desirable to incorporate some form of protection against overload. Many different protection systems are available and each has its own snags and advantages. For the purpose of the following discussion it is assumed that the protection circuit is to cause the output to fall to zero when a certain load current is exceeded.

Figure 8.7 is the circuit of a simple stabilizer in which the inclusion of R_4 and R_8 offers some protection against overload, because T_3

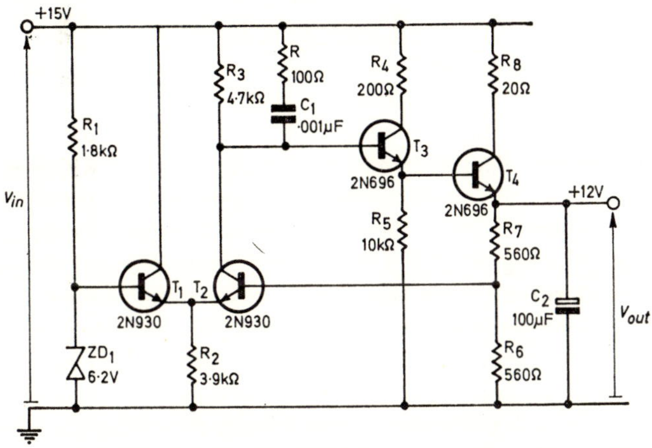

Fig. 8.7 Stabilizer with simple protection by R_4 and R_8

and T_4 bottom at high load currents. Under short-circuit conditions, however, the load current is then approximately V_{in}/R_8 which is likely to be several times the normal load current if V_{in} and the normal value of V_{out} differ only slightly. In the example, R_8 cannot be much larger than 20 Ω because T_4 would then bottom at the normal load current of 100 mA. If the load becomes low resistance then T_4 bottoms and the load current is about 750 mA. Although the stabilizer may survive this, T_4 being safe from the dissipation point of view, the load may be damaged and the source V_{in} overloaded.

An attractive solution is to add a resistor R_0 to the circuit in series with the load current and to cause the output voltage to fall to zero when V_{R0} exceeds a certain level. Since accuracy is unimportant to within about 15 per cent, the V_{be} of a transistor T_5 can be the critical level and when T_5 turns on it can be made to short-circuit the reference Zener diode ZD_1 thus reducing the output to zero (Fig. 8.8).

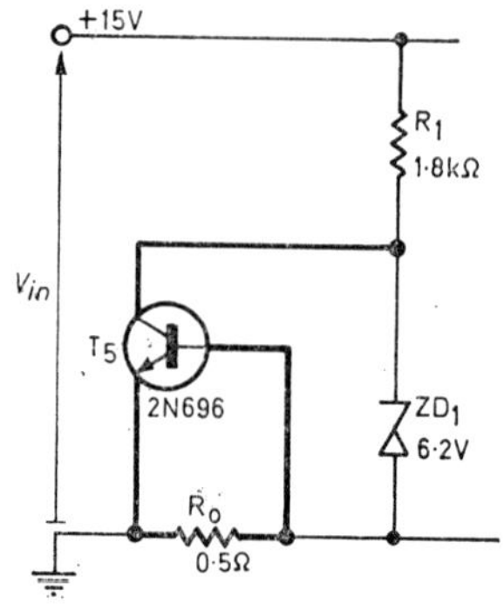

Fig. 8.8 Attempted overload trip

There are two elementary mistakes in this proposal which can be corrected by small circuit changes, and a more fundamental error which considerably limits the usefulness of the idea. Firstly in the circuit given in Fig. 8.8, when the output is short-circuited, the short-circuit load current flows through T_5 base emitter circuit. Unless the protection operates extremely rapidly T_5 will be destroyed before the output is reduced. This fault is easily remedied by inserting a series resistor R_B of a few hundred ohms directly in T_5 base. Secondly, when ZD_1 is short-circuited – it is actually driven slightly negative by the drop across R_0 if T_5 bottoms perfectly – the stabilized output does not necessarily fall to zero. Although T_1 is cut off and T_2 causes the output to fall, T_2 collector cannot quite reach zero because this would imply zero current in R_2; T_2 would then be cut off and could not be driving R_3 towards zero. Because of the V_{be} of T_3 and T_4 it is possible with some circuit values that V_{out} does reach zero but in most cases it will remain at least a few hundred millivolts positive. This is of course unsatisfactory in the presence of a sbort-circuit load; T_4 and perhaps also the load would be destroyed (unless the protection of R_8 is in any case sufficient) by the enormous load current. This snag is overcome by connecting T_5 collector to T_2

collector instead of to ZD_1, the two modifications being sbown in Fig. 8.9.

The remaining problem concerns the basic assumption that V_{out} becomes zero, providing the above details are observed. In fact, if during a short-circuit the output actually did fall to zero, the current

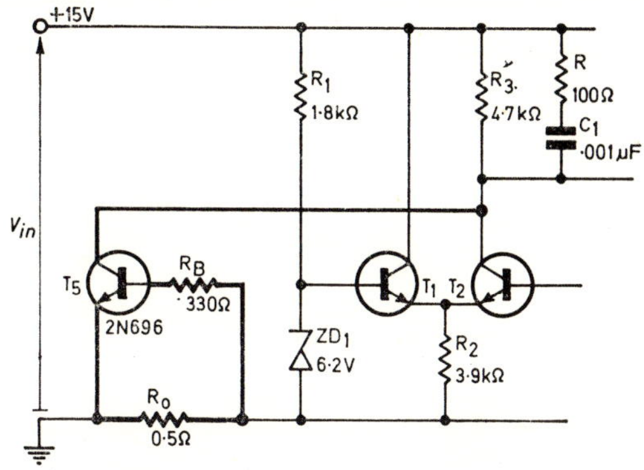

Fig. 8.9 Improved version of Fig. 8.8

in R_0 would disappear, T_5 would be cut off and V_{out} would rise. What really happens is that V_{out} falls until the output load current is the value required to turn on enough current in T_5 to produce the corresponding value of V_{out}. For all practical purposes all that need be said is that the drop across R_0 is V_{be5} when protection is in operation; the load current at short-circuit will therefore be V_{be5}/R_0. Note the (usually) desirable feature that as the temperature rises the sbort-circuit current falls. It is reasonable for load protection to make V_{be}/R_0 slightly greater than normal full load current; when the load resistance falls, V_{out} is then reduced, maintaining the load current at the value V_{be}/R_0.

T_4 is now dissipating considerably more power than under its usual operating condition—it carries just over normal full load current with a collector voltage of V_{in}. Its power dissipation is therefore greater on a short-circuit load by an amount approximating to the power normally taken by the load itself.

This method, therefore, protects the load adequately and limits the current to a known value, but gives very high (though calculable) dissipation in the series transistor under short-circuit load. If such a load is likely to remain for more than a few milliseconds, a suitable heat sink must be provided for T_4. This solution is often adopted where ample volume is available and many commercial bench power supplies use this basic protection system. The pitfall is to assume that V_{out} falls completely to zero thus *reducing* the dissipation in T_4.

Even if the true operation is appreciated there is still at least one more trap when devising a scheme which really does bring V_{out} to zero. One idea is to add another transistor to the trip circuit, forming a complementary bistable (EDH, page 181) as shown in Fig. 8.10. Provided the 'I_{cbo} resistor' R_{10} has a sufficiently low value both

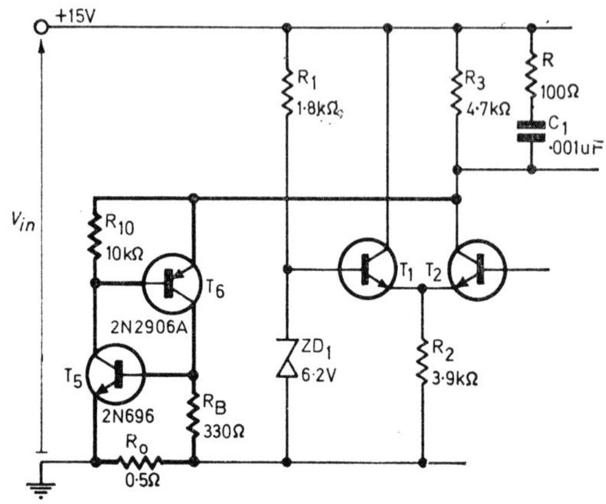

Fig. 8.10 Trip using bistable

transistors will normally be cut off. When a heavy load brings T_5 into conduction, T_6 also turns on and brings T_2 collector voltage towards zero. (Because of the extra V_{be} drop in T_6, T_2 collector is not quite at zero voltage but V_{out} is itself at zero because of the V_{be} drops of T_3 and T_4.) The current now flowing in T_6 is approximately V_{in}/R_3 and this can be designed to hold on T_5 in the absence of any other voltage across R_0, i.e. $V_{in}R_B/R_3 > V_{be5}$. T_6 will then still conduct heavily even if the output load current falls to zero.

In this mode, a short circuit load causes the T_5,T_6 combination to bring the output to zero with virtually no dissipation in T_4. Removal of the short-circuit load leaves the output at zero giving definite indication that an overload has occurred.

When the fault causing the short-circuit load has been cleared it is necessary to reset the protection circuit so that T_5 and T_6 are both cut off. This can sometimes be done by switching the input supply off and on again but this may be inconvenient; an apparently simple method requiring only a low power switch is to short-circuit momentarily the base-emitter junction of either T_5 or T_6, using a spring-loaded push-button or toggle switch (S_2 or S_1, Fig. 8.11).

This is the remaining trap referred to above – the power supply now has the following overload characteristic. On receiving a short circuit or excessive load current the output falls to zero thus protecting both supply and load: on now pressing the reset button, both the

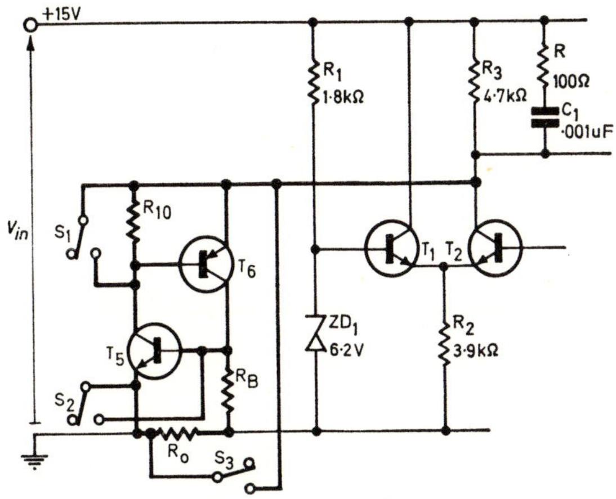

Fig. 8.11 Possible reset for Fig. 8.10 (*see text*)

supply and the load are left unprotected and are destroyed! Only extremely rapid release of the reset switch gives any chance of survival for T_4.

The reason for this behaviour is obvious, yet this trap is fallen into regularly by normally competent designers when devising their first protection circuit. The reset arrangement described unfortunately overrides the influence of output current on T_5,T_6, thus releasing T_2.

A satisfactory reset S_3 consists in short-circuiting momentarily T_6 emitter to T_5 emitter. This cuts off T_6 but holds T_2 collector at zero; on releasing the short-circuit T_6 remains off and allows the output to reach its correct level unless the drop across R_0 becomes excessive as V_{out} rises. In the latter case, indicating that the fault is still present, T_6 turns on again and causes V_{out} to fall to zero (Fig. 8.11). In this way the reset system is safe whether or not the fault has cleared, and whether or not the reset switch is held over for a long time.

PART 2

meeting a specification

CHAPTER 9

the free running multivibrator

FREE running and monostable multivibrator circuits are often the first complete circuits encountered by the student. Their basic action is easily grasped and only an elementary knowledge of transistor operation is required to execute a successful design. These standard circuits suffer, however, from several limitations which make them unsuitable for many applications. In this chapter these limitations are pointed out and remedial measures suggested.

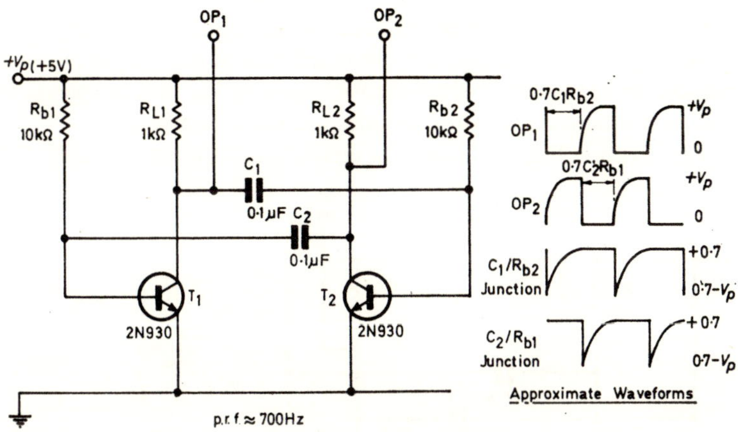

Fig. 9.1 Simple multivibrator

The normal operation of the free running multivibrator circuit (Fig. 9.1) has been described in many text books and will not be repeated here. Circuit properties which often affect performance adversely are detailed below.

reverse base drive

It is obvious when examining circuit operation that when one transistor turns on, the resulting fall of its collector voltage causes the case of the other transistor to fall to approximately $-V_p$; with most modern transistors, the reverse base emitter voltage rating is about 5 V so that V_p should not exceed this figure.

If V_p does exceed the reverse base-emitter rating the result is either the destruction of one or both transistors, or incorrect periodic timing of the circuit. As a rough guide, capacitors exceeding a few hundred picofarads usually cause transistor failure. Lower values result in the negative base swing being cut short at about -7 V, i.e. at the Zener breakdown voltage of the base-emitter junction, giving a faster recovery to the 'on' state and thus a shorter cycling period.

remedies

Four remedies are commonly used; each has its own limitations and advantages. The first is merely to reduce the supply voltage to a safe value, e.g. 5 V; the snags are: reduced output voltage swing and reduced timing accuracy since the variations in V_{be} with temperature and with different transistor samples are more significant.

The second, Fig. 9.2, which is equivalent to returning only the

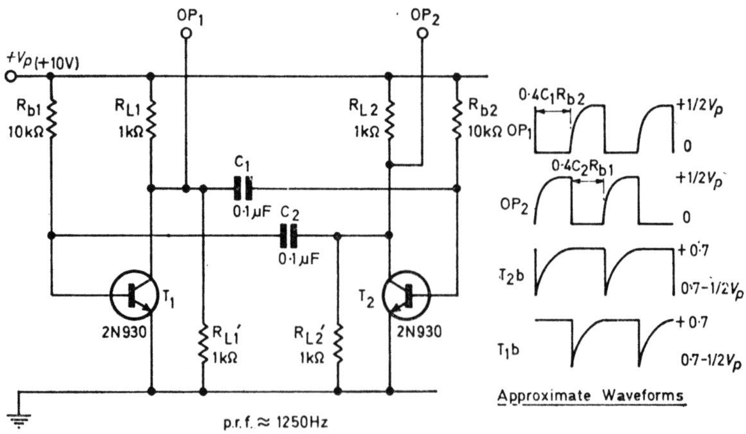

Fig. 9.2 Reducing reverse base drive by limiting collector excursion

collector loads to a lower supply voltage also gives lower output swing but timing accuracy is not quite so badly affected since the charging resistors R_{B1},R_{B2} are still returned to $+V_p$.

The third remedy, Fig. 9.3, is to protect by series diodes so that

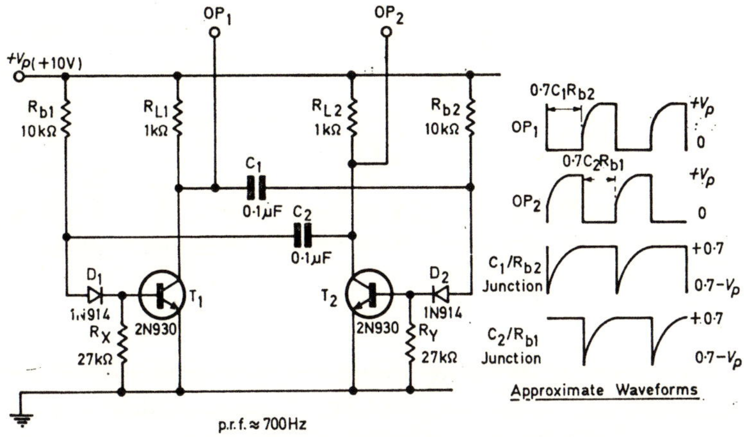

Fig. 9.3 Using base-circuit protection diodes

the base-emitter junction is never reverse biased. Note especially the use of R_x and R_y, without which T_1 and T_2 may not turn off at all (see Chapter 1).

One snag here is that a transistor which has just been in heavy saturation will be slow to turn off unless driven negative by a source that allows reverse base current to flow. This can be overcome if R_x and R_y can be returned to a safe negative supply, e.g. -5 V; the current flowing through these resistors must of course be allowed for when calculating R_{b1} and R_{b2} for the 'on' condition of T_1 and T_2.

Another snag is the extra diode drop which affects the charging time. In this respect, this method begins to show an improvement over the first when V_p exceeds twice the reverse V_{be} rating.

The fourth remedy Fig. 9.4 allows the emitter to descend as soon as reverse current starts to flow in the base-emitter junction.

This does not require resistors since cut off is assured. Speed of cut off is high because of the powerful base drive but timing accuracy is affected as in the previous circuit. The penalties are that, because of the diode voltage drop, the output is smaller, less well defined and does not fall close to zero (a common requirement if another transistor is to be switched, but see below).

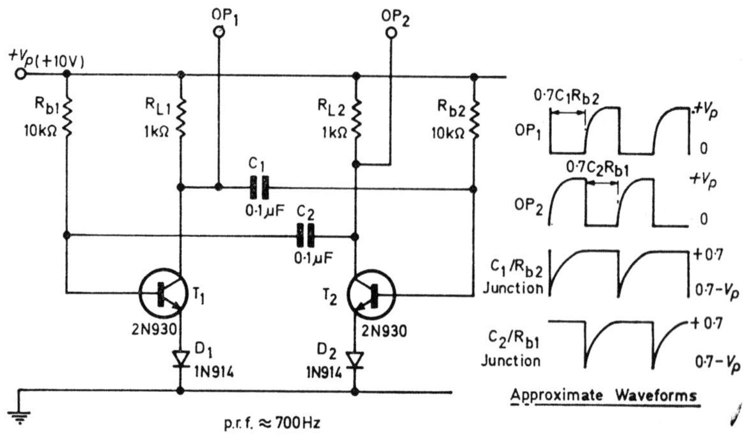

Fig. 9.4 Using emitter-circuit protection diodes

When none of the above remedies is suitable because of the reduced timing accuracy, the possibility of finding a high reverse V_{be} transistor should not be overlooked—many such *p-n-p* alloy types are still available, and a few planar devices, both *p-n-p* and *n-p-n* have ratings of 15 V.

poor collector rise time

Referring to the basic circuit (Fig. 9.1), when T_2 base potential rises to conduction level, T_1 cuts off and its collector rises towards V_p. This rise is noticeably slow and is determined by C_1 and R_{L_1}. Even though the time constant $C_1R_{L_1}$ is normally much less than $C_1R_{B_2}$, the rise time is a considerable proportion of the period because it takes about $4R_{L_1}C_1$ for the collector voltage to rise close to $+V_p$, whilst the base timing curve terminates at about $0{\cdot}7R_{B_2}C_1$. For a design based on $\beta = 40$, one can therefore expect the collector rise time to occupy about 2/7 of the half period.

Again there are several remedies, the effectiveness of which depends on the exact application.

The first, shown in Fig. 9.5, is to isolate the collector circuit from the capacitor during positive swings by means of a diode. The additional resistor R''_{L_1} is essential, and its value is calculated to ensure that the left-hand connection of C_1 reaches $+V_p$ before T_1 turns on

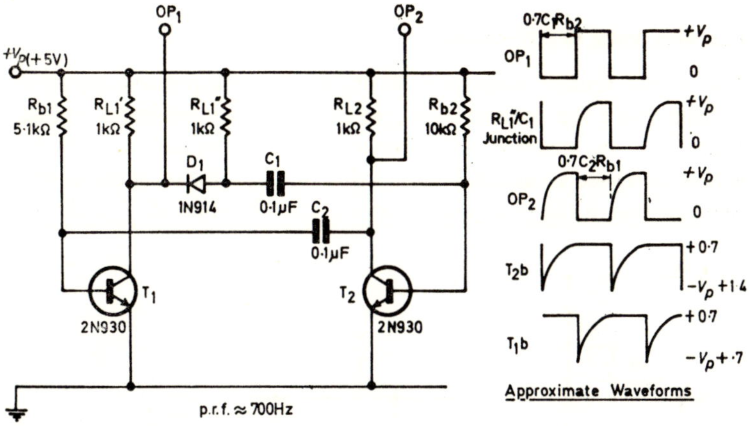

Fig. 9.5 Improved rise time using isolation diode

again. Too large a value means that V_p will not be reached within this time and when T_1 turns on, the excursion at T_2 base will be less than intended, reducing the 'off' time for T_2. This condition implies $4R''_{L_1}C_1 < 0{\cdot}7C_1R_{B2}$, giving R''_{L_1} a similar value to R_{L_1} in the original design. The value of R'_{L_1} depends on the external load: if this is highly capacitive or takes appreciable current to ground, R'_{L_1} may have to be of the same value as R''_{L_1} (to avoid excessive rise time or loss of amplitude); if external loading is light, then it may be possible to have $R'_{L1} \approx 10R''_{L_1}$. In most applications, the nature of the load requires $R'_{L_1} \approx R''_{L_1}$. In such cases R'_{B1} must be halved to drive $R_{L_1}/2$ when T_1 is bottomed, and C_2 doubled to restore the timing. Provided only one output is required (shown here as T_1 collector), there is no need to make the circuit symmetrical and the original values may be used for R_{L_2}, R_{B2} and C_1.

Timing accuracy is impaired since, in the off state of T_1, both terminals of D_1 reach V_p. When T_1 turns on, T_1 collector falls by the diode forward drop before D_1 anode begins to move. This results in a reduced drive to T_2 base, by an amount which varies by 0·6 to 0·9 V with temperature and with the particular diode used, giving reduced off time for T_2.

The philosophy behind this method is to isolate the rising collector voltage from the coupling capacitor. An alternative and fundamentally quite different approach is to force the capacitor to reach its aiming potential more quickly so that this isolation is not required; the second and third remedies are based on this approach.

remedies using fast re-charge

The second remedy consists in returning R_{L_1} to a positive supply of at least twice the voltage of V_p and catching the collector voltage above $+V_p$ by means of a diode (see Fig. 9.6). The new value of R_{L_1} will be designed to give the same value of T_1 collector current when bottomed as in the basic circuit. Since, after T_1 cuts off, T_1 collector

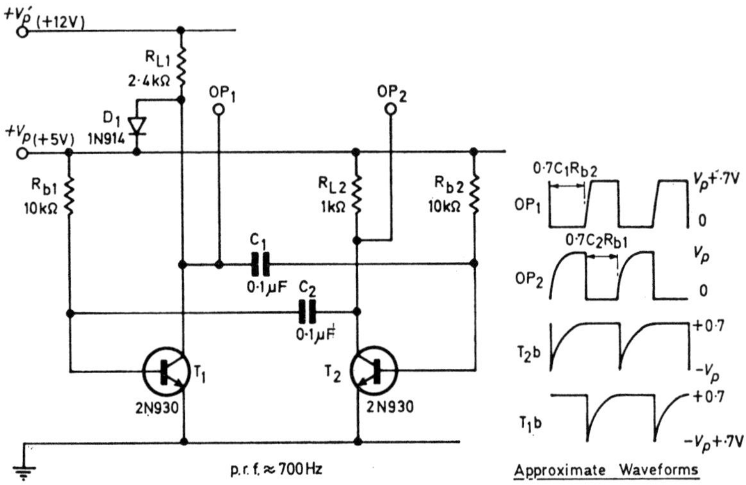

Fig. 9.6 Use of extra supply to recharge C_1 quickly

voltage aims at $+V_p'$, it reaches V_p sooner than in the original circuit, and therefore has a faster rise-time. If this is difficult to appreciate, note that the current which charges C_1 is the same as before, namely the current which has just been flowing in T_1 collector; and the current does not now decrease so rapidly with the rise of T_1 collector voltage. In practice R_{B1} and R_{B2} would also be returned to $+V_p'$ because the variations in V_{be} would then cause less uncertainty in the periodic time.

The obvious snag with this method is the need for an extra supply line; in many large systems several supplies are available, but in small instruments their provision is virtually impossible and this circuit must then be discarded.

The third method, using the same idea as the second, is to add an emitter follower to the circuit so that the charging current for C_1 is increased. Figure 9.7 shows the most usual arrangement, but this is not always the best. It is based on the plausible idea that in order to charge C_1 quickly to $+V_p$ when T_1 cuts off, an *n-p-n* is appropriate since it can pass a large current in the right direction. The simple arrangement shown is however unsatisfactory because T_3 collector current and T_2 base current are virtually unlimited (unless T_3 has controlled β_{max}).

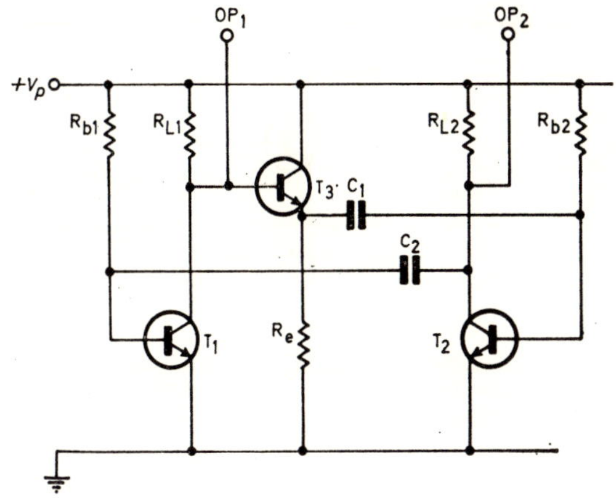

(a)

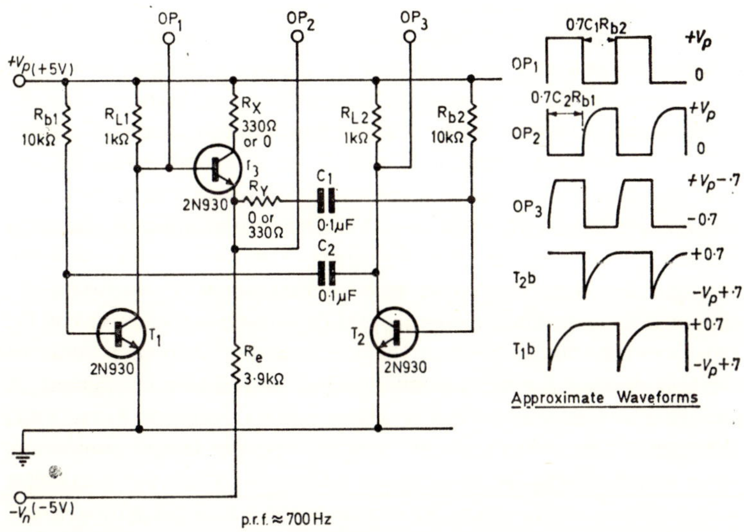

(b)

Fig. 9.7 (*a*) Unsafe circuit for fast recharge of C_1; (*b*) improved version of (*a*)

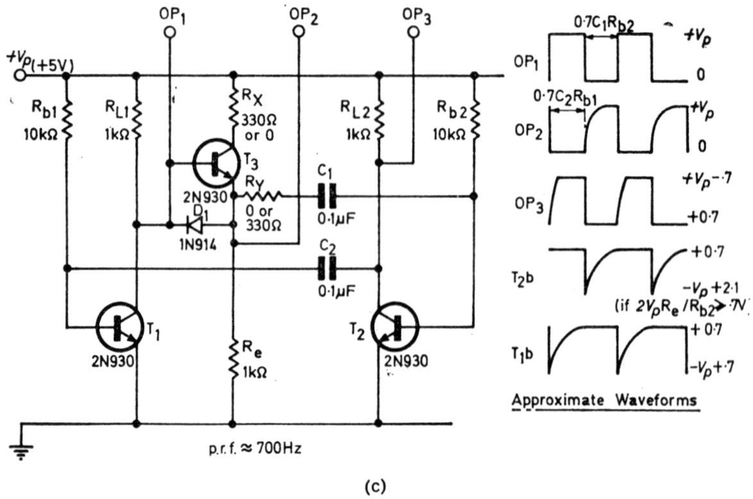

Fig. 9.7c Modified version of (b) with no negative supply

Moreover, when T_1 turns on, T_3 emitter will not reach zero potential because at that instant R_{B2} is attempting to pass a current of about $2V_p/R_{B_2}$ through C_1, thus raising T_3 emitter. The swing on T_2 base is therefore reduced and timing accuracy suffers.

These defects can be overcome by adding a resistor R_x or R_y and returning R_E to a negative supply (see Fig. 9.7b). Either R_x or R_y serves to limit I_{B2} to a maximum value of about V_p/R_x or V_p/R_y; the negative supply $-V_n$ and R_e are designed so that $2V_p/R_B$ flowing through R_e does not cause T_3 emitter to rise higher than $-0{\cdot}7$ V, i.e. $2V_pR_e/R_B < |V_n| - 0{\cdot}7$. The provisioning of the extra negative supply may again be difficult and another solution is given in Fig. 9.7c. Here the diode ensures that when T_1 bottoms, T_3 emitter falls to within a diode drop of the bottoming potential of T_1. This is clearly not so good as the circuit of Fig. 9.7b since more of the voltage step for the capacitor is lost and timing is affected. A preferred solution is to use a *p-n-p* emitter follower (Fig. 9.8); although at first sight the wrong choice, it requires fewer components for a sound design. R_e is designed to supply either the maximum permissible I_b to T_2 when T_1 first cuts off, i.e. V_b/R_e, or a lower current if extreme speed of recharging of C_1 is not required. No extra resistor, like R_x or R_y in Fig. 9.7c is required. There is still a loss of V_{be} in the voltage swing applied to C_1 since when T_1 cuts off, T_3 emitter is not driven above V_p, yet when T_1 bottoms it descends only to $+V_{be}$ above zero. As before, returning R_e to an extra supply line (greater than V_p) would eliminate this error.

In choosing between *n-p-n* and *p-n-p* emitter followers it is useful to examine load conditions. For very light loads OP_1 in either circuit would be used since the greater swing is available; the lower level is very near earth (often important) and in the case of Fig. 9.8 OP_1 will have a faster rise time than OP_3 since it is isolated from C_1.

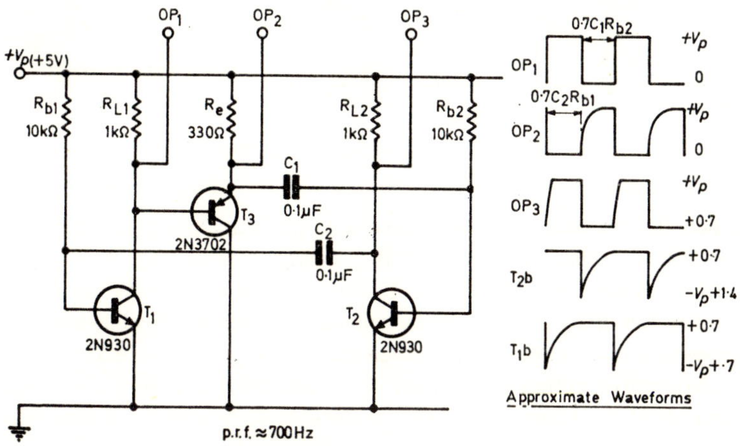

Fig. 9.8 Preferred fast recharge circuit

If the load is highly capacitive, OP_3 must be used in either circuit in order to achieve high charging currents. If R_y rather than R_x is used, then Fig. 9.7c gives a faster rise time than Fig. 9.8 provided the current available from T_3 in Fig. 9.7c is greater than that given by R_e in Fig. 9.8. The latter circuit gives a better fall time since a large base current is available for T_3 from the bottomed T_1.

For this reason it is advisable to add a resistor in series with OP_2 (Fig. 9.8), when driving a heavy capacitive load, to limit the maximum emitter current for T_3 (see Chapter 2).

Another connection for the emitter follower is given in Fig. 9.9; although superficially the least attractive, it can sometimes prove superior to the previous methods. The idea is to reduce R_{L_1} to a value which gives the desired rapid recharge of C_2, comparable with R_e in Fig. 9.8. The extra drive is then obtained from T_3.

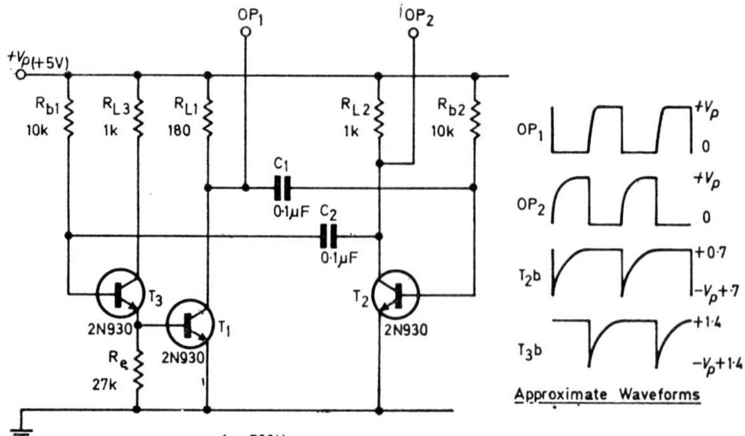

Fig. 9.9 Alternative fast recharge circuit

Since the product $\beta_3\beta_1$ will have a large tolerance it is necessary to limit T_3 collector current by R_{L_3}, designed to be, say, $R_{L_1}\beta_1/2$ so that T_1 is certain to bottom if T_3 bottoms. Alternatively, the direct connection of T_3 collector to T_1 collector is safe; but this has another drawback (see below). R_{B1} is given the value $R_{L_3}\beta_3/2$ so that normal multivibrator action will occur. R_e is essential to ensure the rapid turn off of T_1 when T_3 cuts off; preferably R_e would be returned to a negative line. The main attribute of this rather inelegant looking circuit is that the output from T_1 can drive heavy loads rapidly in either direction with full amplitude (no V_{be} drops). If, however, the collector of T_3 is joined to that of T_1, R_{L_3} being omitted, it now falls only to about 0·7 V, i.e. V_{be} of T_1, not zero, thus losing one of the virtues of the circuit.

The extra transistor T_3 causes degradation in timing accuracy, as in the other circuits used for improving the rising edge, but in this circuit the half cycle affected is the interval when T_1 is cut-off, rather than the 'on' period of T_1.

remedies using output amplification

Yet another approach is to add an output amplifier between the multivibrator and the load, either by tapping off a fast current waveform and converting to a useful voltage swing or simply by voltage amplification and clipping of the poor waveform.

The former is illustrated in Figs 9.10a and 9.10b; use is made of the fact that although T_1 collector voltage is slow to rise, its emitter current (and collector current) drops to zero very rapidly. In Fig.

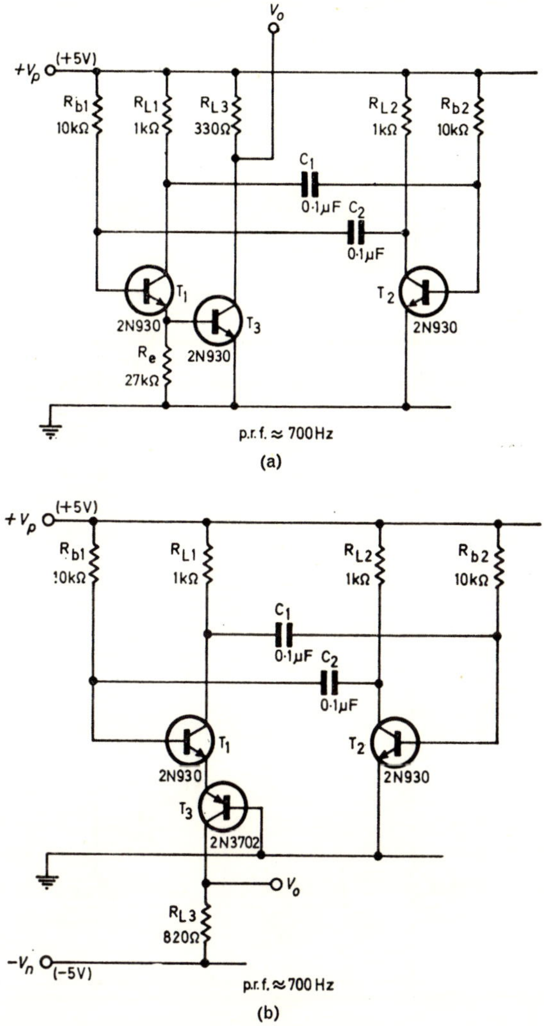

Fig. 9.10 (*a*) Output drive by amplifying T_1 emitter current; (*b*) output drive using earthed base amplifier

9.10a, T_3 base is driven by T_1 emitter current, R_e being used as in Fig. 9.9 to ensure rapid cut-off for T_3. This arrangement is therefore capable of driving heavy loads, although if heavy bottoming of T_3 is allowed the output waveform stays near zero volts for a longer time than T_1 collector owing to the hole storage in T_3. The output voltage is in phase with T_1 collector voltage.

An alternative arrangement (Fig. 9.10b) passes T_1 emitter current

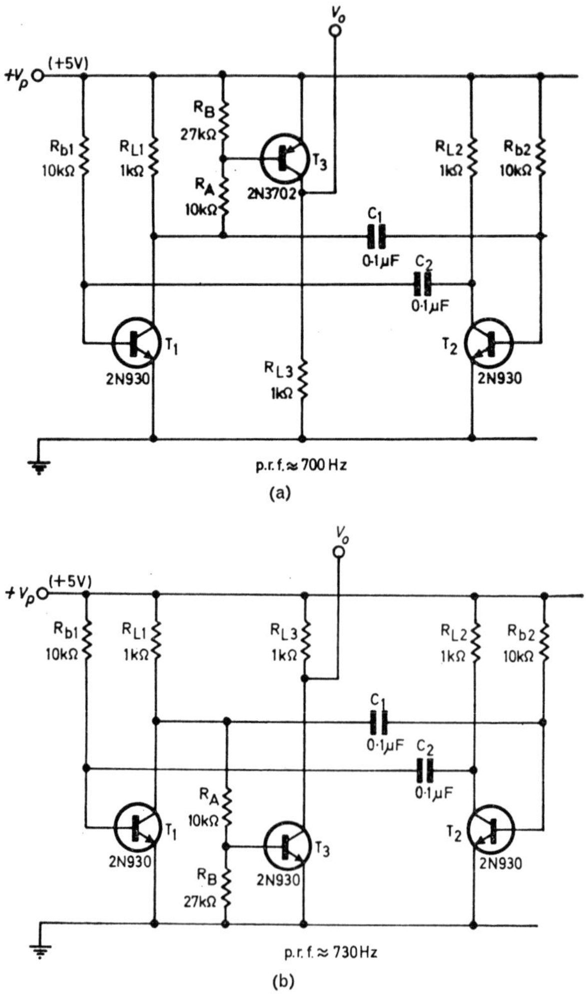

Fig. 9.11 (*a*) Amplifying T_1 collector voltage; (*b*) all *n-p-n* version of (*a*)

into T_3, which has its base earthed. A negative supply, $-V_n$, is required and the output waveform swings between about +0·5 V and $-V_n$, in antiphase with T_1 collector voltage. This circuit is fast if R_{L3} is designed to avoid heavy bottoming but the load driving capability is limited by the operating current used for T_1.

Instead of converting a fast current waveform into a useful voltage, the circuits of Fig. 9.11 merely amplify T_1 collector voltage.

The two circuits give different results, since in Fig. 9.11a the resistors R_A and R_B do not attenuate the collector swing of T_1 unless they are so low as to prevent bottoming. In Fig. 9.11b R_A gives direct attenuation of T_1 collector voltage which now rises to

$$V_p R_A/(R_A + R_{L_1}) + V_{be3} R_{L_1}/(R_A + R_{L_1})$$

instead of V_p. This reduces the interval during which T_2 is cut off and must be allowed for in the timing calculations. Both give an inverted waveform compared with T_1 collector; Figure 9.11a drives heavy loads better when rising, Fig. 9.11b drives harder when falling to zero.

Normally R_A is designed to ensure bottoming of T_3 on full load. If in 9.11a, $R_A + R_{L_1}$ is too high to ensure that T_3 cuts off when $(I_{CO}T_1 + I_{CO}T_2)$ is considered, then R_B is designed to meet this requirement—otherwise, it may be omitted. (Note that the use of a leaky electrolytic capacitor for C_1 may prevent T_3 turning off!) In Fig. 9.11b, R_B must satisfy a similar criterion for $I_{CO}T_3$ but must also satisfy the requirement that $V_{ce(sat)}$ for T_1 after attenuation by $R_B/(R_A + R_B)$ must be less than the threshold for conduction of T_3 (about 0·5 V).

In many ways the circuit of Fig. 9.11a is the best remedy of all the methods described for improving the output rise time. It does not affect the timing of the circuit by adding extra V_{be} drops; changes of external load have no effect on the multivibrator itsclf; within reason, i.e. until T_1 fails to bottom, any number of drivers may be added from T_1 collector, provided separate R_A and R_B resistors are used in each case.

This circuit is, however, little used when the load is light since the simple diode arrangement shown in Fig. 9.5 is often adequate.

It also loses much of its attraction when for other reasons it is essential to recharge the capacitor rapidly; when this has been achieved by one of the circuits shown in Figs 9.6 to 9.9, there is usually no need for an additional speed up circuit to drive the load.

Although all the circuits described have used only one-sided attachment, they may also be used in both halves of the circuit simultaneously.

starting problems

One of the main snags of the standard multivibrator for many applications is its occasional refusal to begin oscillating, although once started, it will continue to operate.

In examining an oscillator circuit for possible failure to start, the technique is to break the a.c. feedback loop without disturbing direct couplings, and to calculate the d.c. conditions which then apply. In non-linear oscillators most of the d.c. levels in this static condition will be different from the intended dynamic values. The loop gain should now be calculated on the assumption that the loop is closed *without causing a transient* (this condition may require that certain capacitors are supposedly pre-charged to a suitable voltage, or that inductors carry an initial current). If the loop gain is then less than unity self-starting is doubtful (it may come about in practice due to a transient) and a redesign may be required.

On applying this criterion to the standard multivibrator of Fig. 9.1, the loop may be broken between C_2 and its connection to T_1 base, leaving R_{B1} connected to T_1 base. Under these conditions T_1 and T_2 would be bottomed in a normal design and it is obvious without actual calculation that if C_2 is reconnected no oscillation will begin, provided C_2 is pre-charged to a voltage $(V_{be1} - V_{CE2})$. This implies that, should this situation ever be reached, even momentarily, the circuit will rest with T_1 and T_2 bottomed.

Although it is interesting to speculate how this could arise the fact that the possibility exists at all could be disastrous in a large system. (One common cause is the slow switch-on of supplies if a mains power unit is used.) Where such a failure has serious consequences it is futile to try to avoid the issue by ensuring fast switch-on of supplies and by providing a starting pulse, because the failure mechanism is non-reversible, i.e. the circuit will not restart, if it ever stops.

The right approach is either to accept the possibility and provide a starting switch which, for example, momentarily short circuits one of the base-emitter junctions, or switches the supply off and on again; or to provide a positive mechanism which restarts the oscillation automatically.

The starting switch may well be adequate where an operator is always available and where failure for a few seconds is unimportant. Hundreds of thousands of multivibrators are in use, relying entirely on this starting method. Where a starting switch is unacceptable, the following suggestion is only one of several methods which ensure self-starting. Other methods can be invented readily when the principle is appreciated, that *there must be loop gain when the loop is closed quietly*, i.e. without a transient.

In Fig. 9.12, each base resistor is returned to its own collector rather than the positive supply. The idea is that each collector must then be clear of bottoming since there is no standing base

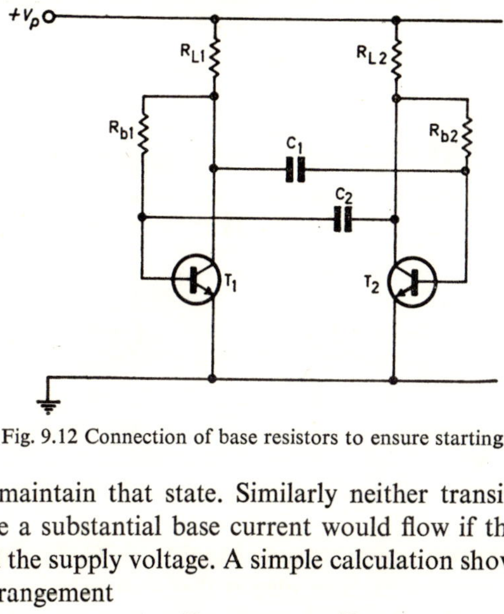

Fig. 9.12 Connection of base resistors to ensure starting

current to maintain that state. Similarly neither transistor can be cut off since a substantial base current would flow if the collectors approached the supply voltage. A simple calculation shows that with this bias arrangement

$$V_c = \frac{\dfrac{V_p}{R_4} + (\beta + 1)\dfrac{V_{BE}}{R_{B1}}}{\dfrac{\beta + 1}{R_{B1}} + \dfrac{1}{R_{L_1}}}$$

giving in typical designs V_c between $V_p/8$ and $V_p/2$.

Both transistors are therefore biased in the linear region and will amplify; with the loop closed oscillation will build up provided this amplification is sufficient, which is assured with any reasonable design.

asymmetrical timing

In familiarizing the student with the principles of multivibrator design it is usual to consider only the symmetrical circuit. In practice, a symmetry is a common requirement and its achievement is worth attention since it is not as simple as it looks. On the other hand, surprisingly large degrees of asymmetry can be produced with the correct technique, so that the usual radical cure of adding a one-shot circuit to the system need not automatically be adopted (but see Chapter 10).

In Fig. 9.13 an attempt is made to obtain a mark/space of about 1/50 from OP_1 by making $C_1 = 50C_2$ on the assumption that the

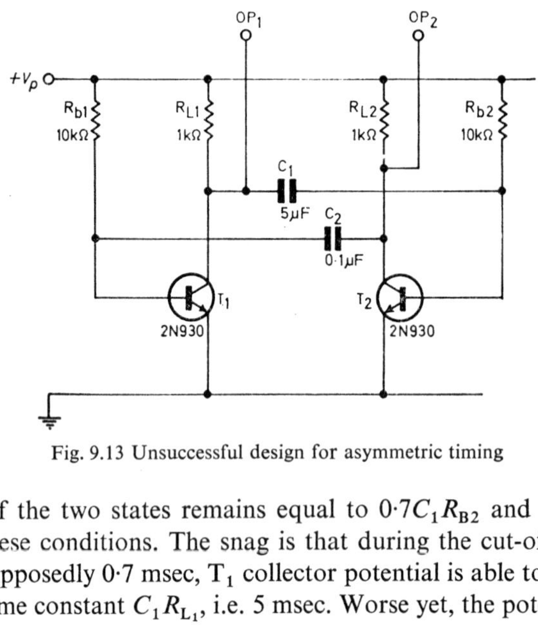

Fig. 9.13 Unsuccessful design for asymmetric timing

timing of the two states remains equal to $0{\cdot}7C_1R_{B2}$ and $0{\cdot}7C_2R_{B1}$ under these conditions. The snag is that during the cut-off interval of T_1, supposedly 0·7 msec, T_1 collector potential is able to rise only on the time constant $C_1R_{L_1}$, i.e. 5 msec. Worse yet, the potential it is intended to reach in 0·7 msec, namely V_p, is the aiming potential itself, so that the time required to achieve it is about $4C_1R_{L_1}$, i.e. 20 msec. During the 0·7 msec interval, T_1 collector potential therefore rises only a small fraction of V_p and when T_1 turns on again the resulting drive to T_2 base is very small. T_2 is therefore cut off for a short time, much less than the intended $0{\cdot}7C_1R_{B2}$ (3·5 msec).

Indeed, if V_p is small, it is quite possible that the drive to T_2 base is insufficient to maintain oscillation and the circuit fails completely.

This problem also appeared when considering collector rise times. Any of the cures mentioned earlier which cause faster recharge (rather than collector isolation from the capacitor or output waveform shaping) are also effective cures here, enabling the large capacitor ratio to be used. These methods do, however, require extra components and if economy is important it is of interest to see how far one can increase the asymmetry by correct design of the standard circuit.

The first requirement is that the ratio of R_{B2} to R_{L_1} must be large since C_1R_{B2} determines the longer interval and $C_1R_{L_1}$ determines the recovery of T_1 collector voltage. From the point of view of R_{L_1} drive, a large value of R_{B2} is in fact advantageous, but R_{B2} must be small enough to bottom T_2.

When T_2 turns on, its collector voltage after only a small descent, causes T_1 to cut off. This provides a large base drive to T_2 through C_1 and R_{L_1}, sufficient to cause T_2 to bottom even if R_{B2} is too large to do so. If T_2 bottoms rapidly, T_1 base reaches $-V_p$ and T_2 collector current at this time is therefore $(2V_p/R_{B1}) + (V_p/R_{L_2})$. (Note that the collector load on T_2 does *not* behave like a load $R_{B1}\|R_{L_1}$ connected to V_{p1} if one is calculating the base current required to bottom T_2.) As T_1 base voltage rises, T_1 collector voltage is also rising and in a circuit which is optimized for asymmetrical operation, T_1 collector voltage just reaches V_p as T_1 turns on. At that time the contribution from $R_{L_1}C_1$ to the base current of T_2 ceases and the load on T_2 collector becomes R_{L_2} as the remote connection of C_2 becomes stationary. The conclusion to be drawn is that in designing R_{B1} and R_{B2} the a.c. coupled loads can be ignored because the base drive is available from the opposite collector load. This is an oversimplification which has a weakness, revealed best by an extreme example.

Supposing, on this theory $R_{B1} = 10\ \text{k}\Omega$, $R_{L_1} = 1\ \text{k}\Omega$, $R_{B2} = 1\ \text{M}\Omega$, $R_{L_2} = 100\ \text{k}\Omega$, C_1/C_2 meeting the condition $4R_{L_1}C_1 = 0{\cdot}7R_{B1}C_2$, i.e. $C_1/C_2 = 1{\cdot}8$, e.g. $C_1 = 0{\cdot}18\ \mu\text{F}$, $C_2 = 0{\cdot}1\ \mu\text{F}$. Now T_1 is readily bottomed by R_{B1} and T_2 is bottomed in the static condition by R_{B2} and in the off period of T_1 by R_{L_1} driving C_1. A time ratio of C_1R_{B2}/C_2R_{B1}, i.e. 180/1, would be predicted by normal theory.

The weak point of the theory is the assumption that when T_2 bottoms, T_1 base is driven to $-V_p$; this is true only if the descent is infinitely rapid. In fact V_{b2} has to move about 0·1 V to cause bottoming with typical values of V_p; this takes about $0{\cdot}1R_{B2}C_1/V_p =$

$C_1 \times 100\ \text{k}\Omega/V_p$ sec, compared with the $R_{B1}C_2$ product of $C_2 \times 10\ \text{k}\Omega = C_1 6\ \text{k}\Omega$. For normal values of V_p these times are comparable and some of the initial excursion on T_1 base is lost.

In the limit it is possible that the current diverted from R_{B1} into T_2 collector, through C_2, during the initial fall of T_2 collector may still leave T_1 bottomed (especially if the designer has been particularly careful in choosing R_{B1}/R_L to be small!). Then, T_1 collector does not rise and no assistance is therefore given by R_{L_1} in accelerating the turn-on of T_2, i.e. no regeneration takes place. The circuit then fails to oscillate.

This is the danger in attempting too great an asymmetry and in practice 20/1 is more realistic provided it is achieved mainly by resistor, not capacitor, unbalance.

combined modifications

Troubles rarely occur singly in any circuit and most multivibrator applications require asymmetrical timing, large and fast output, and protection against reverse V_{be} drive. Fortunately, the modifications described can usually be combined without difficulty if the circuit operation is correctly understood. Fig. 9.14 illustrates such a design in which O.P.1 is 10 V pk, the rising edge being isolated from C_1 by D_1. The cut off time for T_1 is about $0{\cdot}7C_2R_{B1}$, i.e. 17·5 msec. The cut off time for T_2 is about $0{\cdot}5C_1R_{B2}$, i.e. 0·5 msec, since the reverse drive for T_2 is only $V_p/2$. The recovery time for C_1 is about

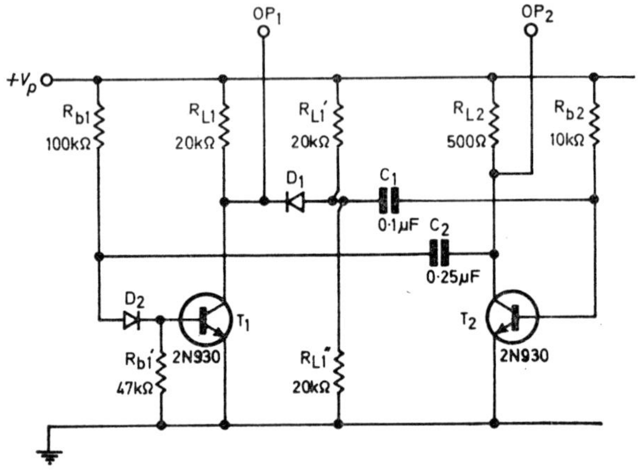

Fig. 9.14 Combination of several modifications

$4C_1R'_{L_1}\|R''_{L_1}$, i.e. 4 msec and for C_2, $4R_{L_2}C_2$, i.e. 0·5 msec. These times are compatible with the reverse base drive periods so the circuit should operate predictably, but it must be remembered that V_{be} drops and leakage currents have been ignored. O.P.1 tends to have a faster edge when rising than when falling, since the latter is dependent upon the rate of rise of T_1 base voltage on the $R_{B1}C_2$ time constant until regeneration takes over.

In conclusion it can be said that the multivibrator although regarded as the most suitable beginner's circuit is far from simple. Its design to meet unusual criteria and its exact analysis can be extremely difficult especially in asymmetrical forms. Fortunately its saturating mode of operation allows bad design errors to be hidden and for most purposes a rough approach gives an adequate design. Most of its bad features are easy to overcome by one or more of the modified forms described.

CHAPTER 10

monostable circuits

THE standard monostable or one-shot is usually taught, like the multivibrator, as being a simple circuit having several common applications. Again, its operation and the consequences of some of its less obvious features are not always made clear and it is relevant to examine these in detail (see Fig. 10.1).

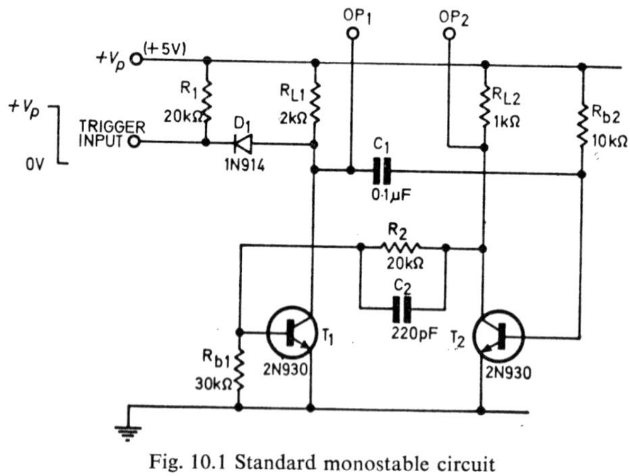

Fig. 10.1 Standard monostable circuit

With no connection to the trigger input, T_2 rests in a bottomed condition, T_1 is cut off and T_1 collector voltage is close to V_p; this assumes a correct design where $R_{B2} < \beta_2 R_{L_2}$. If the trigger input is driven negatively towards zero with a fast edge, the resulting fall in T_1 collector voltage is coupled to T_2 base by C_1, cutting off T_2. T_2 collector voltage rises and the current from R_{L_2} and R_2 causes T_1 to bottom, provided

$$\frac{1}{R_{L_1} + R_2} > \beta\left(\frac{1}{R_{L_1}} + \frac{2}{R_{B2}}\right)$$

Whether or not the input signal is still present, T_1 collector voltage now remains near zero. T_2 base has fallen sharply by approximately V_p and starts to return to its original level as R_{B2} charges C_1, taking about $0{\cdot}7C_1R_{B2}$ sec. Now T_2 turns on again and when it bottoms T_1 is turned off, provided R_{B1} and R_2 are correctly designed.

The output from T_2 collector (OP_2) is therefore a positive-going pulse of duration $0{\cdot}7C_1R_{B2}$ and from T_1 collector (OP_1) a negative-going pulse of the same duration.

triggering problems

There are four points in the circuit where a triggering signal can be applied, each having different snags and advantages.

In the method just described—for most applications the best choice—it is essential that the trigger signal falls sharply enough to cut off T_2 through the coupling capacitor C_1. The most rapid edge is required when T_2 has high β because then C_1 is able to rob T_1 base of a large proportion of its R_{B2} current before T_2 collector begins to rise. Supposing a 'careful' designer makes $R_{B2} = 10R_{L_2}$, allowing a large safety margin in ensuring bottoming even if $\beta \approx 10$. If the actual β is 100, then the base current can fall from V_p/R_{B2} to $V_p/10R_{B2}$ with no appreciable change in T_2 collector voltage. This implies that T_1 collector voltage could be driven negatively at a rate

$$\frac{dV_{C_1}}{dt} = \frac{9}{10}\cdot\frac{V_p}{R_{B2}}\cdot\frac{1}{C_1}$$

and T_2 would just remain bottomed, giving no one-shot action and no output from OP_2.

As a numerical example, if $R_{L_2} = 1\ \text{k}\Omega$, $R_{B2} = 10\ \text{k}\Omega$, $V_p = 5$ V, $C_1 = 0{\cdot}1\ \mu$F, giving an output pulse width of about

$$0{\cdot}7R_{B2}C_1 = 0{\cdot}7 \text{ msec}$$

The trigger rate must not exceed

$$\frac{9}{10}\cdot\frac{5/10^4}{0{\cdot}1 \times 10^{-6}} = 4{\cdot}5 \text{ V/msec}$$

Provided this condition is met, this triggering method has the special advantage that after the initial descent of T_1 collector followed by the continued bottoming of T_1 from $(R_2 + R_{L_2})$, further trigger pulses have no effect on the timing of the final turning on of T_2. This is a most valuable feature and is the main reason why the method is preferred.

Other trigger inputs are shown in Fig. 10.2. Trigger input A has a slight advantage over the original method in that its rate of rise is

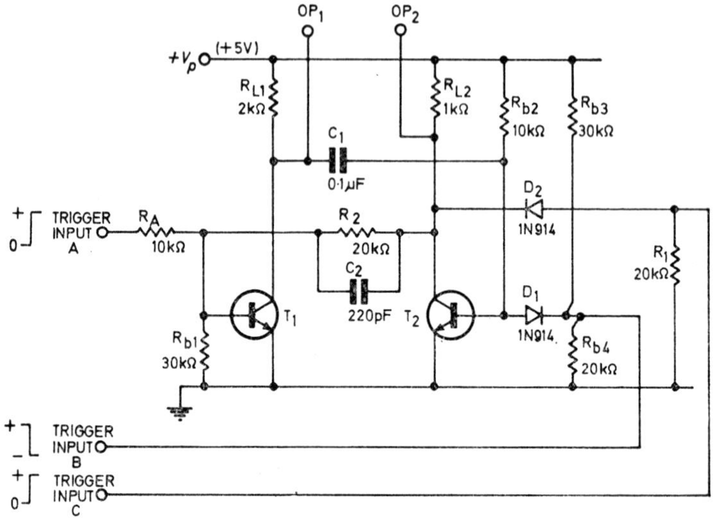

Fig. 10.2 Possible alternative trigger inputs

amplified by T_1 and can therefore be used with a more slowly rising input signal. It has, however, a snag; after successful triggering, a further trigger pulse occurring before T_2 turns on again is coupled through R_A and R_2 to T_2 collector. Since T_1 is bottomed at this time the breakthrough pulse is greatly attenuated by the T_1 base-emitter path but may still represent a spurious output of tens of millivolts especially if R_2 is bypassed by a paralleled speed-up capacitor.

Trigger input B again will operate on an even slower descent, the rate now being amplified by both transistors in a regenerative manner. However, a second trigger pulse occurring just before T_2 is about to turn on again drives T_2 base more negative and delays turn-on. Worse yet, if the trigger signal is a step of longer duration than the intended output pulse width, T_2 will in fact turn off for a total time somewhat longer than the trigger pulse.

This extra time also depends on trigger amplitude. The trigger sensitivity expends on the ratio R_{B3}/R_{B4}; in some designs R_{B3} is omitted, giving very high sensitivity which can be useful or dangerous according to the interference and trigger signal levels present.

A fourth trigger input is just possible; this is T_2 collector which could be driven positive, through a diode (D_2). A point in its favour is that the output edge is then coincident with the trigger edge. However, a trigger of longer duration than $0{\cdot}7R_{B2}C_1$ holds T_2 collector high for just the length of the trigger pulse (T_2 has already turned on and its collector falls as soon as the trigger ends). This also holds T_1 on for the same time, so that no one-shot action occurs. This effect can be useful but there are simpler methods to obtain it. The overwhelming disadvantage of this trigger mode is the large drive required; in a circuit design with $R_{L_2}/R_{B2} = 10$ and an actual β_2 of 100, a drive of 100 mA would be needed if $V_p = 5$, $R_{L_2} = 500$, i.e. much more than R_{L_2} itself passes.

For most applications, T_1 collector triggering as in Fig. 10.1 is best. When the polarity required for this is inconvenient, method A (Fig. 10.2) is often acceptable.

If the output of the one-shot circuit is taken from T_2 collector, which is usual because of the better waveform (see next section) the trigger pulse duration appears to be unimportant if applied to T_1 base or collector.

Whether the trigger pulse is shorter or longer than the output pulse width seems not to affect the timing except for very short triggers, which may not be 'seen' by the circuit. Although this is basically true, there are some side effects if a long trigger pulse is used.

Towards the end of the output pulse, when T_2 base is rising and T_2 collector falls, T_1 base is driven towards earth. If at this time the trigger pulse has ended (or in the case of trigger input A, is sufficiently small), T_1 collector rises and assists the turn-on of T_2. The output therefore falls rapidly due to this regenerative action. If, however, the trigger is still present and of sufficient amplitude to hold T_1 collector at a low potential, T_2 turns on at a rate determined by $R_{B2}C_1$. When the output pulse width is less than a few microseconds, $R_{B2}C_1$ is small, T_2 turns on quickly and the output fall time is limited by f_T and stray capacitance. But when a long output pulse is defined, e.g. 1 msec, the rise time at T_2 base multiplied by the gain in T_2 is still a few microseconds as seen at the output from T_2 collector.

Thus, a trigger pulse longer than the defined output pulse width results in a poor fall time from the output for pulses longer than a few hundred microseconds. This is often important when the final edge is to be used for triggering another circuit, or where it is differentiated to give, say, a 1 μsec pulse. The pulse could then be lost; a cure is effected by differentiating the input trigger pulse.

T_1 collector rise time

T_1 collector circuit behaves exactly like the corresponding multivibrator circuit (Chapter 9) in that its recovery time is about $4R_{L_1}C_1$. As in the multivibrator this gives a poor waveshape from OP_1 and gives the wrong pulse timing if recovery has not occurred before the next trigger pulse is applied. If the OP_1 pulse shape is important but circuit recovery time is not, i.e. the interval between trigger pulses is several times the output pulse width, then the isolation or shaping modifications given in Figs 9.5, 9.10 or 9.11 may be used. If circuit recovery is also important then C_1 must be recharged rapidly and the modifications of Figs 9.6, 9.7b or 9.7c, 9.8 or 9.9 may be used.

possible simplifications

It is evident that the standard one-shot circuit has several features, not all of which may be required in every possible application. For instance, the trigger pulse need only be long enough to start the action; its amplitude has no influence on circuit performance over a wide range of amplitudes. Two complementary outputs are available, OP_2 having very fast edges due to regeneration (unless trigger duration exceeds output pulse width).

In many applications the possibility of a simpler circuit with lower cost and probably higher reliability can be considered. A circuit which can often replace the one-shot is shown in Fig. 10.3; it could be designated a 'half-shot' circuit as it operates in the same way as the T_2 half of the one-shot circuit of Fig. 10.2. For its successful use, the input must have a known constant amplitude of duration greater than the desired output pulse. The output pulse is inverted in direction and is not assisted in final turn-on by any regenerative action. Within these limitations the circuit often substitutes for a one-shot.

One very common application is in conjunction with a multi-

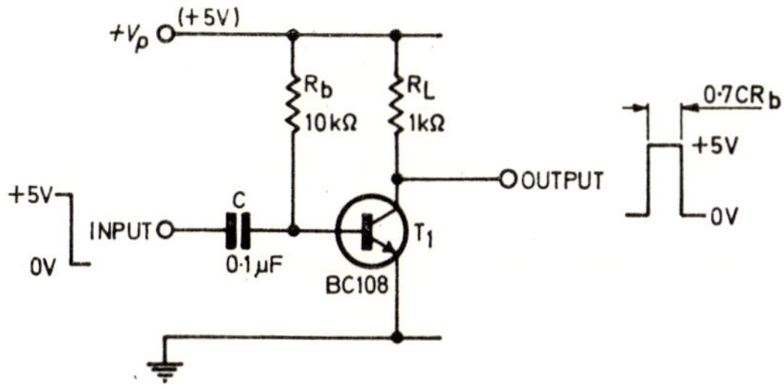

Fig. 10.3 'Half-shot' circuit

vibrator when a very short mark/space pulse waveform is required (see Chapter 9). When the requirement is too stringent for direct generation it is easy to use a symmetrical multivibrator directly connected to the 'half-shot'. The driving signal is ideal, being long in duration and defined by cut-off and bottoming levels; bonus is given in the isolation provided by the 'half-shot' in absolutely removing any influence of the load on the cycle time.

This is one of many cases where it is rewarding to think carefully before using a circuit that has more features than are really required. Simplification is achieved with no loss of reliability; in fact the simpler circuit, unlike the one-shot, is virtually immune to triggering from spurious interfering spikes and is therefore safer in practice.

other modifications

Almost all the modifications described for the multivibrator in Chapter 9 are also applicable here, including combinations such as Fig. 10.4. Here, the reverse rating for $T_2 V_{be}$ is respected by the use of R'_{L_1}, and R''_{L_1}, but a large output is still available from OP_1. This is also quite fast, being isolated from C_1. Since fast recharge has not been provided for C_1, recovery is still slow.

There are certain cases where slow recovery is highly desirable. This occurs in burglar-alarm systems and in arrangements designed to limit mean operating power. For example, in high-current earth-loop testing, a 100 A, 50 V a.c. output may be required for 1 sec during the test. It is uneconomic to make a mains transformer of this continuous rating and it is reasonable to limit the test rate to give a 10 sec interval between tests.

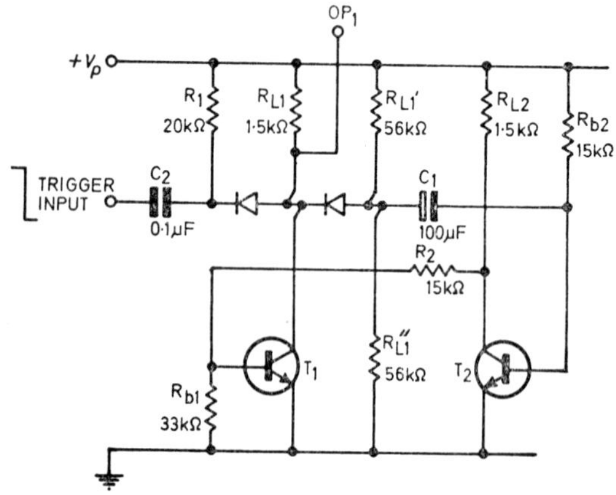

Fig. 10.4 Long recovery one shot

A simple method is to design a one-shot as in Fig. 10.4, with large and equal values of R'_{L_1} and R''_{L_1}. If the output pulse time is designed to be 1 sec and the recovery time $4(R'_{L_1} \| R''_{L_1})C_1$ is 10 sec, then the output pulse can drive a relay which switches on the transformer primary. On triggering the one-shot, a 1 sec test is performed and if the operator triggers again within 10 sec the resulting test pulse is less than 1 sec. The output duty ratio is therefore self-limiting whether or not the operator triggers too soon.

CHAPTER 11

schmitt trigger circuits

THE Schmitt trigger is the name originally applied to a particular form of trigger circuit using two thermionic valves. It has come to mean any form of circuit in which a step change in output occurs when an input, however slowly it moves, crosses a certain direct voltage level; and a step change in output in the opposite direction occurs when the input crosses a different voltage level in the opposite direction.

To illustrate this operation, the circuit of Fig. 11.1 shows the transistor equivalent of the original valve circuit. The action is most

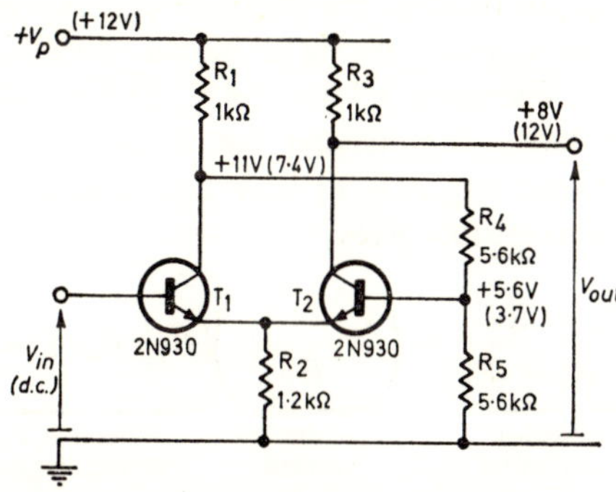

Fig. 11.1 Schmitt trigger

easily examined by assuming that V_{in} rises gradually from zero until triggering occurs and then descends until the circuit triggers back.

When V_{in} is at zero volts, T_1 is cut off so that T_1 collector and T_2 base potential are determined by V_p, R_1, R_4 and R_5; with the values shown this gives approximately $T_{1C} = 11$ V, $T_{2B} = 5{\cdot}6$ V, as indicated by the unbracketed voltage labels in Fig. 11.1. Because T_2 base is at 5·6 V, T_2 emitter will be at about 4·9 V and T_2 emitter current $4{\cdot}9/R_2$, or 4 mA. This current flows in R_3 and T_2 collector rests at 8 V.

As V_{in} rises no change occurs in these levels until T_1 begins to conduct at 5·6 V, i.e. 0·7 V above T_1 emitter potential. When T_1 conducts, its collector voltage falls, causing T_2 base to fall. This reduces T_2's share of the current in R_2 and therefore turns on more current in T_1. This continues until T_1 is taking all the current passed by R_2 so that T_1 collector is at 7·4 V, T_2 base is at 3·7 V, T_2 is now cut off and its collector voltage is at 12 V. Owing to the positive feedback mechanism which controls the transitions, the rate of change of V_{out} is very rapid. It is limited by stray capacitance and f_T but not, to a first order, by the rate of rise of V_{in}.

If V_{in} now continues to rise, T_1 conducts harder and T_2 base falls as T_1 collector potential falls. Eventually T_1 bottoms (at about 6·8 V input) and T_2 still remains cut off.

When V_{in} is now made to fall towards zero, T_1 takes less current and T_2 base rises; when V_{in} has reached 5·6 V on its way towards zero, T_2 base is at 3·7 V as explained above. As V_{in} falls below 5·6 V, T_2 base is still rising and T_1 finally begins to cut off at the point where V_{in} just falls below T_2 base potential. If T_2 base were stationary during this process, the trigger level in this direction would be 3·7 V. However T_2 base is rising all the time and the actual trigger point therefore lies between 3·7 and 5·6 V. The calculation to determine this level looks complicated but is easily understood. The relationship between V_{in} and T_2 base is calculated assuming that only T_1 is conducting; this gives

$$V_{B2} = \left\{ \underbrace{\frac{V_p(R_4 + R_5)}{R_1 + R_4 + R_5}}_{\substack{\text{T}_1\text{ collector} \\ \text{`off' potential}}} - \underbrace{\left(\frac{V_{in} - V_{be}}{R_2}\right)}_{\substack{\text{T}_1\text{ collector} \\ \text{current}}} \underbrace{[R_1 \| (R_4 + R_5)]}_{\substack{\text{Effective T}_1 \\ \text{collector load}}} \right\} \underbrace{\frac{R_5}{R_4 + R_5}}_{\substack{\text{Potential} \\ \text{divider, T}_1 \\ \text{collector to} \\ \text{T}_2\text{ base}}}$$

The value of V_{in} when triggering takes place to cause T_2 to conduct is equal to this value of V_{B2}. Therefore, putting $V_{in} = V_{B2}$ in the above yields V_{in} in terms of the resistors and V_p; in this circuit the lower trigger level is 4·2 V.

In principle this circuit is therefore useful in producing a fast transition when its input signal reaches a certain direct voltage level. The circuit has 'backlash' (a mechanical term) in that the trigger level is different when reverting to the original conditions. Although at times a nuisance, backlash is usually desirable as it ensures that a slowly rising signal, with a fast interfering signal superimposed, triggers only once when rising or falling provided the Schmitt backlash voltage exceeds the peak-to-peak interference.

This circuit is very widely used but suffers from a few disadvantages discussed below.

disadvantages of standard circuits

Although the trigger levels are, with good design, largely independent of transistor parameters, they do depend on several resistor values as shown above. For accurate trigger levels, all resistors except R_3 must therefore be close-tolerance types.

Moreover, the calculation of suitable sets of resistors to give required trigger levels is much more difficult (if standard values are to be used) than the reverse process of calculating the trigger levels from known circuit values.

Another undesirable feature of the standard circuit is the low input impedance seen by the source when the input d.c. level has risen beyond triggering level sufficiently to allow T_1 to bottom. The input impedance which then approximates to $R_1 \| R_2 \| (R_4 + R_5)$ can seriously affect the input waveform to the detriment of another part of the system. In a typical design this occurs at an output level which exceeds the rising trigger level by only a few volts.

A further deficiency for many applications is the small output voltage swing, limited to the difference between V_p and the upper, i.e. rising, trigger level.

remedies

Clearly, some of the above-mentioned snags can be overcome by adding extra stages of buffering or amplification, and the addition of a high resistor in series with the input alleviates the change of input impedance which occurs as T_1 bottoms, although at the expense of trigger level accuracy.

When an extra transistor is definitely required then the three-transistor circuit of Fig. 11.2 will usually prove to be the optimum

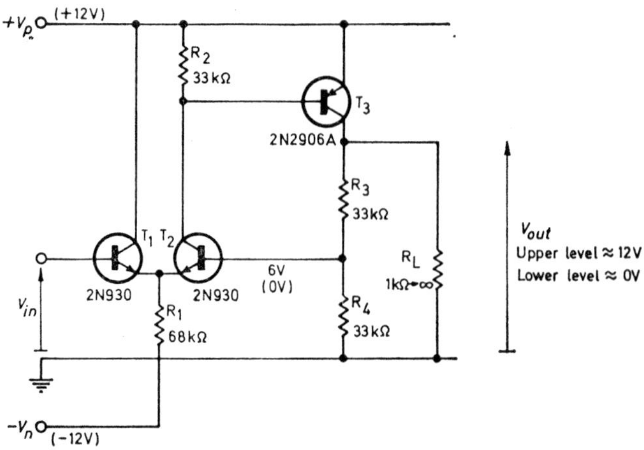

Fig. 11.2 Improved Schmitt trigger

configuration. It is shown here in its most commonly used form where one trigger level is 0 V and the other is positive.

Circuit operation is similar in principle to Fig. 11.1 but voltage levels are easily defined. When V_{in} is small, T_1 is cut off and the current in R_1 is designed to be sufficient to bottom T_3. T_2 base therefore takes up a potential $V_p R_4/(R_3 + R_4)$ (neglecting $V_{ce(sat)}$ for T_3) and since T_1 cannot turn on until V_{in} reaches this voltage, this is the rising trigger level.

When T_1 base exceeds the trigger level, T_2 cuts off and causes T_3 to cut off (provided R_2 is small enough – see Chapter 1). T_3 collector and T_2 base potential both fall to zero and this is therefore the descending trigger level.

Note that to a first approximation the presence of an external load from T_3 collector to earth has no effect on either trigger level provided T_3 still bottoms, i.e. provided

$$\frac{1}{R_1}\left(V_n + \frac{V_p R_4}{R_3 + R_4}\right) > \frac{R_L \| (R_3 + R_4)}{\beta_3} + \frac{V_{eb3}}{R_2}$$

Input impedance, except during the transition, is always at least $\beta_1 R_1$ until T_1 itself bottoms with an input just exceeding V_p; output swing is almost V_p; 0 V and $V_p R_4/(R_3 + R_4)$ V are simply the trigger levels, so that only two resistor values are important in one trigger level and none at all are significant in the other.

In this circuit, as in the original, the trigger levels are also dependent upon transistor β and I_{CBO}; the 0 V trigger level is in fact

$$+(I_{CBO2} + I_{CBO3})R_4$$

in the worst case, i.e. $R_L \to \infty$, and the higher level is depressed by $I_{B2}R_3 \| R_4$. By correct design of operating currents and use of appropriate transistor types these can be made negligible in either circuit.

The circuit of Fig. 11.2 therefore overcomes all the basic snags of Fig. 11.1 in one step: its outstanding disadvantage – often completely unimportant – is that T_3 is driven into bottoming and therefore is slow to cut off. With modern transistors this circuit can still operate at tens of Megahertz but is not to be recommended, for example, as a start or stop gate in a 100 MHz counter-timer. For such high-speed operation, T_3 could be prevented from bottoming by including an emitter resistor and by designing R_1 appropriately. However, some of the virtues of the circuit are then lost – output is reduced and the upper trigger level is less well defined.

other forms of the three-transistor Schmitt trigger

Sometimes zero voltage triggering is not required and then the optimum connection for obtaining the correct levels depends on the nature of the load R_L.

If R_L is quite low compared with $(R_3 + R_4)$, the addition of R_5 as shown in Fig. 11.3 raises both trigger levels; the lower one is now

$$V_P \frac{R_4 \| (R_3 + R_2)}{R_5\ R_4 \| (R_3 + R_2)}$$

and the upper one is

$$V_P \frac{R_4}{R_4 + R_3 \| R_5}$$

The negative supply is not required provided the upper trigger level exceeds about 2 V. Note that the output is slightly reduced because the current through R_5 and R_3 into R_L prevents V_{out} reaching zero on its negative excursion; and that the lower trigger level changes appreciably when R_L becomes large.

If on the other hand R_L is comparable with $(R_3 + R_4)$ but is constant and known accurately, the connection of R_5 to T_3 collector

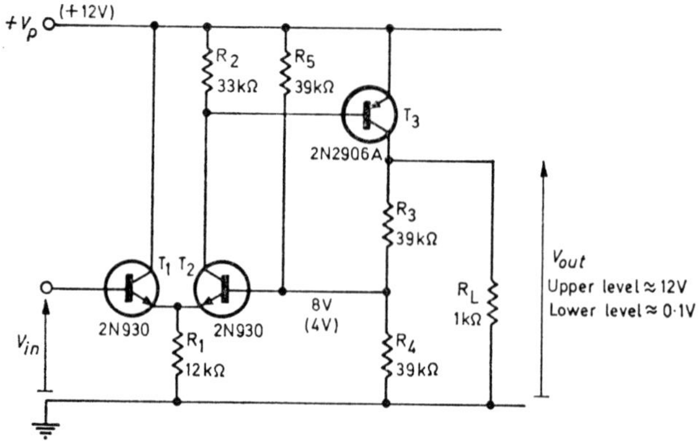

Fig. 11.3 Raising the trigger levels

(see Fig. 11.4) has the advantage that the upper trigger level has the same value as in Fig. 11.2 and depends only on V_p, R_3 and R_4. The lower trigger level is now

$$V_p \frac{R_4}{R_3 + R_4} \cdot \frac{R_L \| (R_3 + R_4)}{R_5 + R_4 \| (R_3 + R_4)}$$

by inspection. The disadvantage compared with Fig. 11.3 is, however, that the output swing is reduced very much more and on its negative excursion only reaches a voltage $(R_3 + R_4)/R_4$ of the lower trigger level.

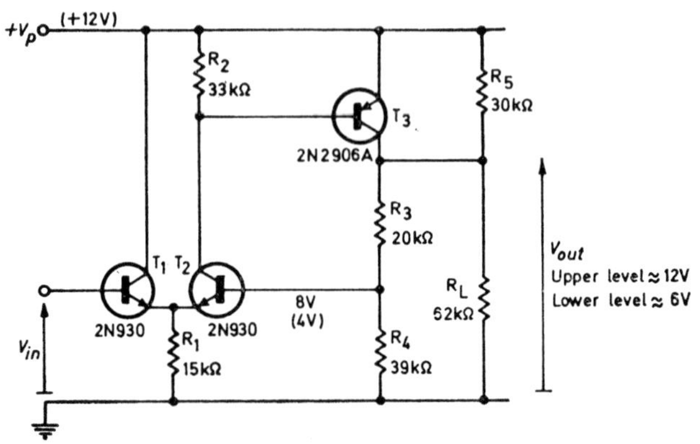

Fig. 11.4 Raising the lower trigger levels

Where the lower output level has to be definitely zero or negative, the use of an isolating diode (Fig. 11.5) ensures that the load swing cannot influence the trigger levels and vice versa, provided the load return is more negative than the lower trigger level.

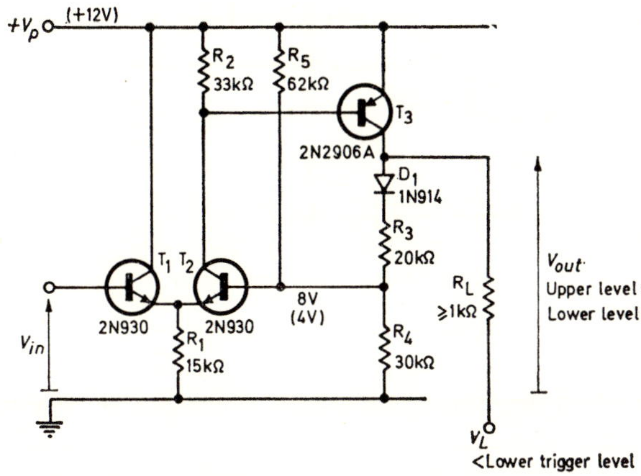

Fig. 11.5 Use of isolating diode (1)

The snags this time are the temperature-dependent uncertainty introduced by the diode in the upper trigger level and the dependence of the upper trigger level on the three resistors R_3, R_4 and R_5.

A slight variation is to connect R_5 as shown in Fig. 11.6. This makes the upper trigger level independent of R_5 and allows the load return potential to be higher; however, temperature variations of the diode forward drop have slightly more influence on the upper trigger level (in the examples given, diode variations are attenuated 2:1 in Fig. 11.5 and 3 : 2 in Fig. 11.6).

The differences in the circuits of Figs 11.5 and 11.6 are small and either version is especially useful when the load voltage is required to fall below earth; V_L can be as negative as the diode and T_3 rating will allow.

refinements

One or other form of the three-transistor Schmitt trigger will meet most requirements in practice. When improvement is needed this

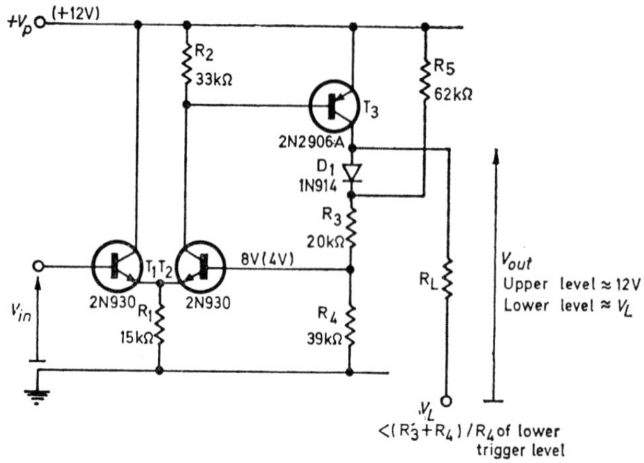

Fig. 11.6 Use of isolating diode (2)

usually involves trigger-level accuracies which can be improved by careful design.

The previous discussion has made several assumptions in order to present a clear picture of the improvements given by the three-transistor circuit. One of these is the inference that triggering begins when the input base reaches the same voltage as T_2 base. This is approximately true provided the V_{be} of T_1 and T_2 are matched, but even then some finite change in input must take place before triggering is complete. This error is less in the recommended circuit than in a normal Schmitt trigger owing to the extra gain provided by the third transistor. The major error in trigger level is normally the difference between the absolute values of the two V_{be}'s and their temperature coefficients. For many designs an adjustment can be provided either on the input or by making part of R_3, R_4 or R_5 variable. When temperature effects are significant it is advisable to use a matched pair for T_1,T_2, preferably a true 'dual' with two matched transistors in the same can.

Most of the remaining error is caused by the initial tolerance and temperature coefficient of R_3, R_4 and R_5. In a design requiring precise trigger-level definition a thin-film network may be used in which ratios of resistance (as required here) will track accurately with temperature and can have initial tolerances of 0·1 per cent.

When a very heavy load, such as a relay or lamp, has to be driven, T_1,T_2 may have to operate at such a current level that their base

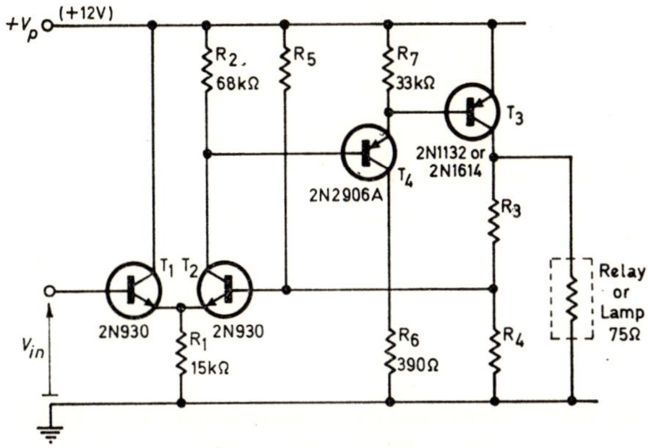

Fig. 11.7 Driving a heavy load

currents affect the trigger levels. This effect can be made negligible in most cases by the use of an extra transistor driving T_3 (see Fig. 11.7). T_4 should be protected from excess dissipation when T_2 turns on, either by R_6 as shown, or by returning T_4 collector to T_3 collector. In the latter case T_4 becomes almost bottomed when turned on; but T_3 collector can now rise only to $(V_p - V_{be3})$ and this affects the upper trigger level.

CHAPTER 12

ramp generation

THERE are many circuit configurations which will provide a ramp voltage waveform followed by a rapid return to the original level. The well-known Miller and bootstrap arrangements can be designed to give adequate performance for most practical purposes. When a particularly fast recovery (flyback) time is required, the complementary bistable (see EDH, p. 181) is often used as a discharge switch because the regenerative action is very powerful, the transistors working in effect as cascaded earthed emitter amplifiers.

In the form shown in Fig. 12.1 the charging circuit produces an exponential rather than a linear rise; this may be linearized by replacing R_1 by a constant current source or by bootstrapping part of R_1 (Figs 12.2 and 12.3).

Although this is a commonly used circuit, its design is difficult to

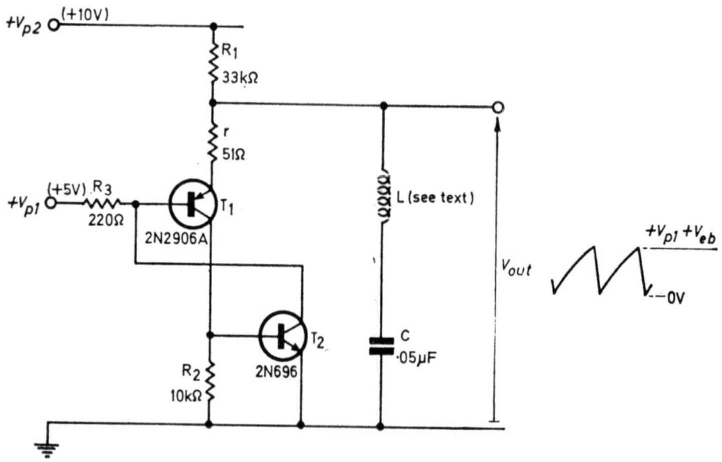

Fig. 12.1 Ramp generator may depend on presence of inductance if R_1/R_2 is badly chosen

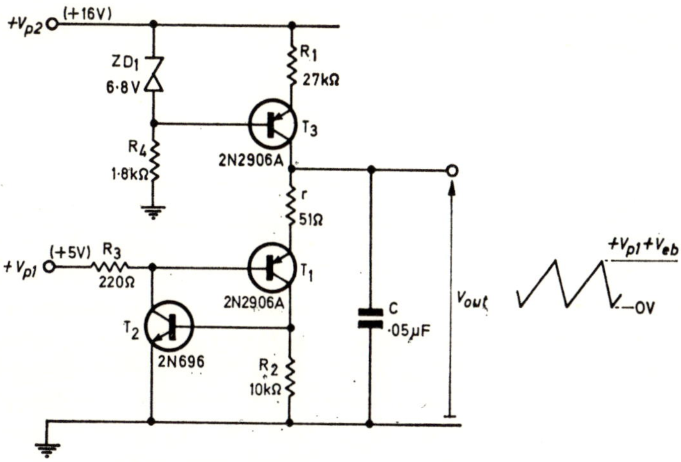

Fig. 12.2 Constant current form of circuit of Fig. 12.1

tolerance for reliable operation. This is particularly true if the p.r.f. is to be varied by voltage control (base of T_3 in Fig. 12.2) or by resistor value (R_1); the practical result is that satisfactory performance is obtained only over a narrow range.

The following description shows up the design problem as a

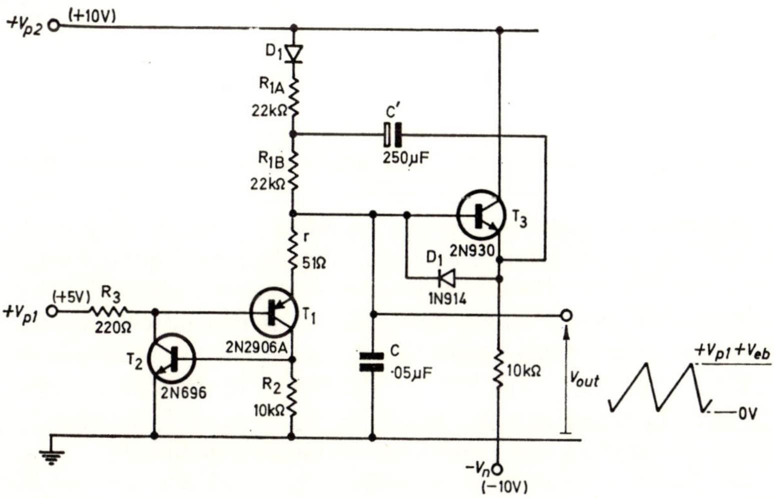

Fig. 12.3 Bootstrap form of circuit of Fig. 12.1

balance between achieving flyback, ensuring recovery from flyback and reasonably low peak current.

circuit operation

The simple version shown in Fig. 12.1 is intended to operate as follows, assuming $L = 0$. At switch-on, C is uncharged, T_1 emitter is at zero volts and T_1 base is at $+V_{p1}$. T_1 is therefore cut off so that the drop across R_2 is small and T_2 is also cut off. C charges from R_1 on a time constant CR_1, aiming at $+V_{p2}$. When T_1 emitter voltage reaches $V_{p1} + V_{eb}$, T_1 conducts. With correct design T_1 collector current is sufficient to turn on T_2 which in turn drives T_1 base towards zero voltage. As T_1 base falls, T_1 emitter potential follows and the emitter current increases very rapidly as C discharges. This turns on T_2 still more until T_1 emitter voltage is close to zero. C cannot now be discharged further, and provided the current now passing through R_1 is small enough, the drop across R_2 is insufficient to hold T_2 in conduction. T_2 cuts off and T_1 base potential rises rapidly to $+V_{p1}$, cutting off T_1 since its emitter is held by C. This completes the cycle and C now begins to charge as before. The emitter resistor r serves merely to limit the maximum discharge current to a safe value.

design

The basic design is straightforward. The output moves from approximately $+V_{be}$ to $V_{p1} + V_{be}$, and travels on a time constant R_1C towards V_{p2}. The maximum current flowing into T_2 base occurs just after the onset of regeneration when T_1 emitter may reach $+V_{be}$ above zero before C has appreciably discharged, giving a surge current of V_{p1}/r. These considerations enable reasonable values for R_1, r, C, V_{p1} and V_{p2} to be determined. R_3 is more difficult to specify; too large a value means that T_1 emitter has to rise noticeably higher than $V_{p1} + V_{eb}$, i.e. T_1 base has to be dragged upwards, before T_1 passes enough current into R_2 to cause T_2 to conduct. This would give a higher and undefined output peak than $V_{p1} + V_{eb}$, because the drop across R_3 depends on β_1. Too low a value for R_3 means that T_2 collector current has to be very high to reach bottoming, possibly to the extent that T_1 emitter current is insufficient to bring about this state. T_1 emitter would not then approach zero potential.

R_2 has similar constraints; too large a value prevents T_2 from turning off when V_{out} approaches zero, the current in R_1 being sufficient to keep T_2 bottomed without the extra current flow produced as C discharges. If R_2 is extremely large, leakage from T_1 and T_2 can turn T_2 on even at switch-on when T_1 emitter is near zero. In this case C does not even begin to charge and no oscillation occurs. Too low a value prevents T_2 turning on so that the regenerative discharge action does not take place.

The result of these conflicting requirements is that, although the basic operating principle is simple, the detailed design of this circuit demands great care. One condition is that T_2 must have insufficient drive to remain bottomed when V_{out} approaches zero and C no longer contributes to T_1 emitter current. This may be expressed mathematically:

$$\underbrace{\beta_{2\max}\left(\underbrace{\underset{\substack{\uparrow\\ \text{Available}\\ \text{current}\\ \text{from } T_1\\ \text{emitter}}}{\frac{V_{p2}-V_{eb1}}{R_1}} - \underset{\substack{\uparrow\\ \text{Current}\\ \text{lost in}\\ R_2 \text{ when}\\ T_2 \text{ is `on'}}}{\frac{V_{eb2}}{R_2}}}_{\text{Base current into } T_2}\right)}_{T_2 \text{ collector current}} < \underbrace{\underset{\substack{\uparrow\\ \text{Current}\\ \text{in } R_3\\ \text{if } T_2\\ \text{bottoms}}}{\frac{V_{p1}}{R_3}} + \underset{\substack{\uparrow\\ \text{Current in}\\ T_1 \text{ base}\\ \text{if } T_2\\ \text{bottoms}}}{\frac{V_{p2}-V_{eb1}}{\beta R_1}}}_{\substack{T_2 \text{ collector current}\\ \text{when bottomed}}} \qquad (12.1)$$

Using the values given in Fig. 12.1, the result expressed by the above equation can be arrived at as follows. If T_1 and T_2 are almost bottomed, T_1 emitter current is

$$\frac{10 - 0{\cdot}7}{33} = 0{\cdot}28 \text{ mA}$$

Of this, $0{\cdot}7/10 = 0{\cdot}07$ mA is taken by R_2, leaving 0·21 mA in T_2 base. If T_2 were to be bottomed by this drive its collector current would be $5/220$ A $= 22{\cdot}7$ mA into R_3 plus $0{\cdot}28/\beta_1$ mA into T_1 base. Therefore, bottoming from this cause (which is to be avoided as it represents the lock-up condition) is most likely if β_1 is high, e.g. greater than 100, giving T_2 collector current greater than 22·7 mA. Since the drive into T_2 base is 0·21 mA the circuit is safe if $\beta_2 < 22{\cdot}7/0{\cdot}21$, namely 108.

The circuit given should therefore operate with the transistors suggested but an attempt to double the p.r.f. by halving R_1 would clearly require a lower value of R_3 also; the lower limit of R_3 is determined by the ability of T_2 to bottom in its presence when driven by the discharge current from C.

Another condition is that the drive to T_2 when V_{out} has exceeded V_{p1} by V_{eb1} must be sufficient to cause regeneration; if not, V_{out} remains at $V_{p1} + V_{eb1}$. (This is obvious if the limit cases $R_2 = 0$ or $R_3 = 0$ are considered.) At this time the drive to T_2, although apparently required to be greater than in the previously considered state, tends to be less because R_1 drops a lower voltage. The saving clause is that T_2 is not now required to bottom, only to conduct sufficiently to ensure that the loop gain with C present is greater than unity to start the regenerative action. T_2 then has to bottom but has the discharge current of C as base drive.

Note that if R_1 is replaced by a constant current source or bootstrap the above conditions are easier to meet, quite apart from other advantages.

When V_{out} has risen above V_{p1} causing T_1 to pass an emitter current I_e, then

$$\Bigg[\underbrace{V_{p2} - \bigg(V_{eb1} + \underbrace{V_{p1} + R_3\frac{I_{e1}}{\beta_1}}_{\text{T}_1\text{ base potential}}\bigg)}_{\text{T}_1\text{ emitter potential across R}_1}\Bigg]\Bigg/R_1 = I_{e1} \tag{12.2}$$

giving

$$I_{e1}\left(1 + \frac{R_3}{\beta_1 R_1}\right) = \frac{V_{p2} - (V_{eb1} + V_{p1})}{R_1}$$

One condition for R_2 (not yet proved to be the most stringent) is that T_2 must at any rate be turned on, i.e.

$$I_{e1}R_2 > V_{be2}$$

or

$$V_{be2}\left(1 + \frac{R_3}{\beta_1 R_1}\right) \leqslant \frac{R_2}{R_1}[V_{p2} - (V_{eb1} + V_{p1})] \tag{12.3}$$

Normally the f_T of T_1 and T_2 is such that C represents virtually zero reactance at the effective frequency at which regeneration takes place; if this were not so, C would by implication be low and definition of frequency would be poor due to stray capacitance seriously affecting the timing. Provided C does appear to be zero reactance, T_2 collector current shares between R_3 and the input impedance of T_1 base, namely $\beta_1[r + (1/g_{m1})]$. It hardly requires calculation to prove that if R_3 is at least comparable with $\beta_1[r + (1/g_{m1})]$, regeneration is assured because the loop contains two cascaded earthed emitter amplifiers.

For completeness, the simplified equations for incremental changes in i_{c2} read as follows. The i_{c1} which occurs from a change in i_{c2} is

$$i_{c1} = \beta_1 i_{b1} = \frac{i_{c2}\beta_1 R_3}{R_3 + \beta_1\left(r + \dfrac{1}{g_{m1}}\right)}$$

The i_{c2} which results from this change in i_{c1} is

$$\beta_2 i_{b2} = \frac{\beta_2 R_2 i_{c1}}{R_2 + \beta_2/g_{m2}}$$

$$= \frac{i_{c2}\beta_2 R_2 \beta_1 R_3}{\left(R_2 + \dfrac{\beta_2}{g_{m2}}\right)\left[R_3 + \beta_1\left(r + \dfrac{1}{g_{m1}}\right)\right]}$$

Regeneration occurs if this exceeds the original i_{c2}, i.e.

$$\beta_2 R_2 \beta_1 R_3 > \left(R_2 + \frac{\beta_2}{g_{m2}}\right)\left[R_3 + \beta_1\left(r + \frac{1}{g_{m1}}\right)\right] \qquad (12.4)$$

Thus R_2 and R_3 may each be as low as $1/(\beta - 1)$ of the associated base input impedance before regeneration fails.

Again referring to the values in Fig. 12.1, when V_{out} reaches 5·7 V, T_1 emitter current is (10 − 5·7)/33 = 0·13 mA; with T_2 conducting, 0·7/10 = 0·07 mA is passed by R_2, leaving T_2 base current equal to 0·06 mA. T_2 is therefore conducting by a reasonable safety margin, i.e. normal tolerances in V_{p1}, V_{p2}, V_{be}, R_1 and R_2 cannot reduce T_2 base current to zero. At the onset of regeneration, taking $\beta_1 = \beta_2 = 30$, T_1 emitter current is 0·13 mA, giving $g_{m1} = 1/350$, and T_2 emitter current is (0·06 × 30) mA = 1·8 mA, giving $g_{m2} \approx 1/30$. Therefore

$$\beta_2 R_2 \beta_1 R_3 \approx 2 \times 10^9$$

and

$$\left(R_2 + \frac{\beta_2}{g_{m2}}\right)\left[R_3 + \beta_1\left(r + \frac{1}{g_{m1}}\right)\right] \approx 0{\cdot}13 \times 10^9$$

This confirms [Equation (12.4)] that loop gain is adequate and the circuit will operate.

The weaknesses of this circuit shown up by the earlier analysis [Equation (12.1)] are the high currents in T_2, heavy current drain on V_{p1} and completely uncertain operation if $\beta_{2\,max}$ is unspecified at a collector current of V_{p2}/R_2. The required value of R_3 is invariably much lower than other considerations would give and if the mode of operation is misunderstood a theoretically inoperable circuit is the result.

'Normal' evaluation of R_3 would merely ensure that

$$(I_{b1} - I_{CBO1\,max} - I_{CBO2})R_3$$

is an insignificant voltage in comparison with both $(V_{p2} - V_{p1})$, for accuracy in definition of charging current, and V_{p1}, for accuracy in the output voltage just before C discharges. In the given example $I_{b1\,max}$ would be about $(10 - 5)/33\beta_{min}$, i.e. 7·5 μA if $\beta_{min} = 20$. Even if the resulting drop across R_3 is allowed to represent only 1 per cent of 5 V, R_3 can be as high as $(5 \times 10^6)/(100 \times 7{\cdot}5)$, namely 6·6 KΩ. From previous reckoning this value would cause the output voltage to sit permanently near zero.

In spite of all these calculations many circuits have been shown to work even though R_3 is in theory much too large. The reason is that an inductance L appears in series with C due to imperfection in the capacitor. Assuming that at switch-on T_1 base rapidly reaches V_{p1}, C charges from R_1 and eventually reaches $V_{p1} + V_{eb1}$. This begins the regenerative action described earlier and V_{out} falls towards earth as C discharges through T_1 collector circuit to T_2 base circuit. With R_3 too large the action would now end if no inductance were present; however, the discharge current has been passing through L and this cannot stop instantaneously so that T_1 emitter continues to fall—driven now by the stored energy in L rather than its own base drive. As soon as T_1 emitter falls below 0 V, T_1 and T_2 are cut off; T_1 base rises to V_{p1} and T_1 emitter begins to rise again since the tuned circuit formed by L and C is now very heavily damped by series resistor R_1. The cycle is now repeated and the circuit operates more or less satisfactorily.

The snags with this mode of operation are that L is unknown, and the Q of the LC circuit in series with the bottomed transistor T_1 and the base circuit of T_2 is also unknown. This gives an unpredictable margin of operation and an output which swings below zero voltage by an unknown amount. Furthermore, the circuit is not guaranteed to start if, for instance, V_{p1} is applied slowly and later than V_{p2}, since the d.c. bottoming conditions discussed earlier then prevent the rise of T_1 emitter voltage.

Some of these problems can be overcome by deliberately adding a known inductance in series with a high quality capacitor and using a sufficiently large r in series to define Q within reasonable limits. However, the starting problem remains, the output waveform becomes badly distorted from the intended simple exponential and the timing depends on L and r.

improvements

As shown above, the successful operation of this circuit is so dependent upon d.c. conditions that a continuously variable frequency-generator of this type can be made to cover only a small range of frequencies.

In most wide-range applications an entirely different configuration would be considered, on the lines discussed in Chapter 6, for example. In seeking a solution to make the present circuit more easily designed, however, an approach comes to light which as a by-product makes possible incredibly low operating frequencies.

The basic weakness of the circuit of Fig. 12.1 is the need to ensure that T_2 cannot bottom when the static value of T_1 emitter current is highest and yet to be certain that T_2 conducts sufficiently for regeneration when T_1 emitter current is lowest. The design procedure outlined above brings these conditions about by reducing R_3 to a value which T_2 is incapable of driving fully except during the discharge.

An alternative approach is to allow R_3 to have a 'normal' value, thus avoiding heavy current demand from T_2 and V_{p1}, and to use a low value of T_1 emitter current (high R_1) so that in the static condition the drop across R_2 cannot turn on T_2 at all. Under these conditions C will charge until T_1 turns on at an output voltage $V_{p1} + V_{eb1}$ and will remain in this state.

Although the d.c. conditions prevent T_2 conduction the dynamic situation is unchanged in that momentary turning on of T_2 still results in a regenerative action which terminates by discharging C. This suggests that a positive pulse applied to T_2 base, when C has reached its full charge, will effect the recycling of the charge circuit.

To produce such a pulse only at this instant would require detection circuitry to monitor the voltage present on C and would be expensive to execute. A train of pulses continuously applied to T_2 base is successful provided the amplification of T_2 is closely controlled. In that case the pulses may be such that their amplified level at T_1 base drives T_1 into conduction only when C is almost fully charged. However, normal variations in $T_2\beta$ cause great uncertainty in the output level at which these pulses manage to initiate the discharge.

A practical realization of the technique is to use instead a continuous train of negative pulses to T_1 base. If a timing and output amplitude accuracy of, say, 5 per cent is required, then the pulses may be 2 per cent of V_p in amplitude, and of p.r.f. about 50 times the ramp p.r.f.

When the ramp is ascending, these pulses have insufficient amplitude to cause T_1 to conduct; as T_1 emitter and base potentials become comparable the pulses bring conduction nearer until finally one pulse initiates the discharge. The ramp timing is slightly affected because unless the pulse train is synchronized, a time uncertainty equal to the pulse spacing exists. The current T_2 required to complete the regenerative loop is provided by release of energy from C caused by T_1 conduction.

This somewhat elaborate method of making this form of ramp generator designable requires two extra transistors to form a pulse source (or a single transistor, plus transformer) and is seemingly unattractive. There are, however, hidden advantages which make the system invaluable in many applications.

The first point is that the final output is synchronized to the pulse waveform. If synchronization to an existing waveform is essential, it is usually a simple matter to derive the trigger pulse from this waveform, thus saving the two transistors.

The second, major, point is that by this method the upper limit value of R_1 has become, by simple theory, infinite. In other words any value of R_1, however great, will always eventually bring T_1 emitter to $V_p + V_{eb1}$, and discharge will occur from the pulse drive. With practical components this is not entirely true because capacitor leakage, I_{EBO1}, leakage of circuit boards and availability of reliable high value resistors combine to limit R_1 to a few megohms.

All but I_{EBO1} may be made virtually as insignificant as cost will allow; for high-temperature use, transistor leakage is, however, very restricting.

A great improvement can be made in this respect by adding an ultra-low leakage diode in series with T_1 emitter. (Diodes in this category are two or more orders better than I_{EBO} leakage.) The snag is a greater variability in ramp output timing and amplitude by an extra -2 mV/degC, due to the diode forward-drop temperature coefficient.

If a high-quality capacitor, e.g. Mylar dielectric, is used, and special mounting precautions are taken in the charging circuit (glazed ceramic insulating pillars) it now becomes possible to make R_1 as high as several hundred megohms.

The ramp time, without resorting to electrolytic capacitors of doubtful internal leakage, can now be of several minutes duration with an accuracy which is normally available only with much shorter intervals.

This technique often allows analogue methods to be used for timing sequences where the only reliable alternative is a much more expensive digital chain.

Figure 12.4 illustrates a slow ramp generator using this principle. C is 6 μF, being a reasonable package size in polyester dielectric, having very low leakage and a low price. V_C at triggering level is about 6·4 V so that for V_{p2} = 10 V, the time for C to charge from

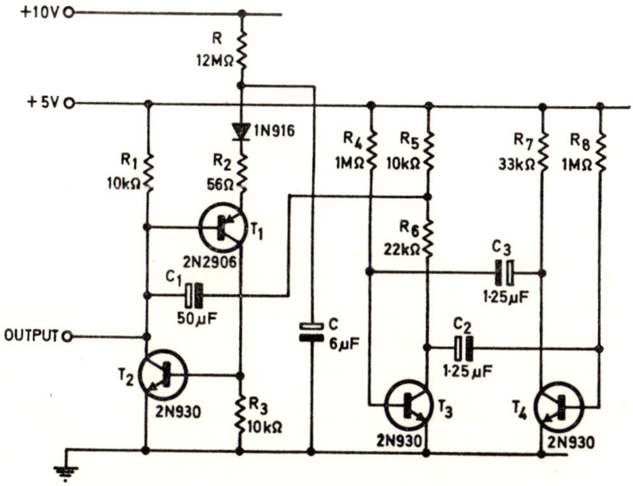

Fig. 12.4 Low speed pulse-generator using ramp timing

zero is approximately *CR*; this gives 72 sec when $R = 12\ \text{M}\Omega$. The pulse generator is a standard multivibrator giving negative pulses on T_1 base with an interval of 1·7 sec; the accuracy of the pulse generator is of only minor significance and electrolytics capacitors are adequate. The only tricky point in the design is to ensure that the pulse arriving at T_1 base causes T_1 emitter to descend quickly enough and far enough – it must cause C to deliver enough current to turn on T_2.

With the values shown, a safe criterion is that T_2 should bottom, requiring 0·5 mA collector current, i.e. about 25 μA base current maximum. R_2 itself requires 70 μA so that C must deliver about 100 μA into T_1 emitter when the pulse arrives. The rate of descent of the pulse must therefore exceed $100/C$ V/sec, i.e. 15 V/sec. The multivibrator has a base waveform rising at a rate of about 4 V/sec at the point of conduction; this is amplified to give a collector fall rate of about 100 V/sec. The pulse coupled to T_1 base is therefore of adequate rate of descent.

CHAPTER 13

feedback amplifiers

NEGATIVE feedback theory and practice has been discussed in many text books and the advantages which result from its use are well known. In designing an amplifier to a particular specification for use with negative feedback there seems to be an infinite choice of configurations, each offering certain advantages and snags. Many circuits which are suitable for a.c. applications have too much drift for d.c. purposes; some are unsuitable for cascading one to the next because input and output d.c. levels differ—for the same reason a gain change brought about by the use of a different resistor value alters d.c. operating conditions; input impedance may be ill-defined.

There is much to be said, therefore, in favour of a circuit which is suitable for a.c. or d.c. use, has easily defined input resistance, allows the gain to be changed without upsetting operating levels, can be cascaded as required, and is not affected by large supply voltage changes.

Several microcircuits are available, e.g. μA 709 which meet these conditions, but also suffer from practical snags. In satisfying a large variety of needs, these circuits often have very high internal gain and very low drift. This results in the need for two sets of external components in order to guarantee freedom from oscillation because of the high gain; moreover, lock-up is a very real problem often requiring an extra transistor to prevent it. In many applications the high gain is required, in which case such a circuit is economic. Often, however, an internal gain of only a few hundred is needed with moderate drift performance. It is then useful to have available a circuit configuration in which only a few easily calculated values need to be changed to suit most of these applications.

basic circuit

The circuit of Fig. 13.1 is the basic configuration which in many cases is sufficient as it stands for practical use. Like all d.c.-coupled high gain amplifiers its exploitation generally requires a negative feedback connection from the OP to the inverting input as indicated by R_5.

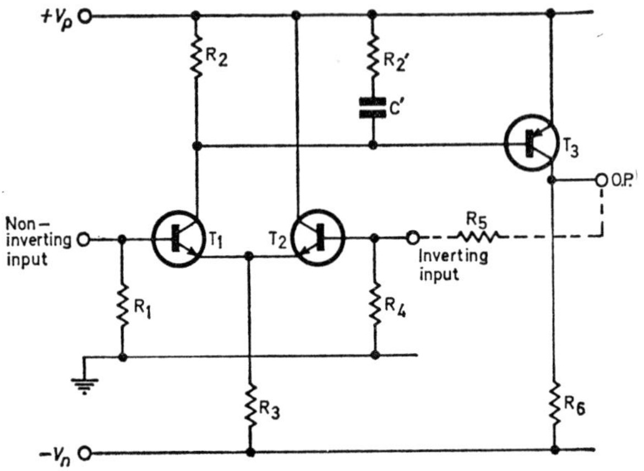

Fig. 13.1 Basic amplifier circuit

Assuming the circuit values are designed correctly as explained later, the base currents of T_1 and T_2 are small enough to cause negligible voltage drop across R_1 and R_4. With no input, T_1 base is therefore at zero potential.

To determine the quiescent potential of T_2 base a *reductio ad absurdum* method can be applied as follows. If T_2 base is positive this implies, with perfectly balanced transistors, that T_2 is conducting and T_1 cut off. This causes T_3 to cut off and there is no path by which T_2 base can remain positive. Similar if T_2 base is negative, T_1 turns on, causing T_3 to conduct sufficiently (with correct design) to drive T_2 base positive. Since neither of these is a stable condition it is evident that the only possibility is that T_2 base will settle at 0 V provided T_1 and T_2 have the same V_{be} and have a very high gain g_m.

Since T_2 base current is assumed to be negligible and there is no voltage drop across R_4, the current in R_5 is zero, and T_3 collector also sits at 0 V.

By similar reasoning, if the V_{be} of T_1 and T_2 differ by δV_{be} then

T_3 collector settles at $V_{be}(R_5 + R_4)/R_4$ with the same polarity as T_2 base potential.

On applying an input signal, T_1 base potential moves so that T_1 tends to increase or decrease its operating current. As described above for the quiescent condition, the result is that T_2 base follows the T_1 base signal exactly. For input e, either a.c. or d.c., T_2 base therefore carries an identical signal e. This implies a current in R_4 equal to e/R_4 and this must be the result of a signal coupled from T_3 collector by R_5 (there is no other possibility, as T_2 base current is negligible). The output signal at T_3 collector is therefore $+e(R_5 + R_4)/R_4$, giving an over-all gain to a.c. or d.c. signals of $(R_5 + R_4)/R_4$.

The above simplified calculation made several assumptions, one of which was the equality of the V_{be} of T_1 and T_2 or at least the implication that any inequality would remain fixed in the presence of a varying input signal. If this were strictly true it would mean that the output signal was obtained without changing the differential signal applied to T_1 and T_2, i.e. gain from T_1–T_2 base to output would be infinite. In practice the gain is typically 500 and the over-all gain is therefore lower than $(R_5 + R_4)/R_4$ by about 0·2 per cent.

The assumption of zero base current in T_1 and T_2 had to be made in order that T_2 base potential could be determined by V_{out} attenuated only by the R_4,R_5 network, and in order that T_1 base potential would be zero with no input. In a real design the two base currents would have to meet the condition that $R_1 I_{B1}$ and $I_{B2}(R_5 \| R_4)$ must each be a negligible proportion of the input voltage. The operating currents of T_1 and T_2, determined by V_n and R_3 must therefore be designed to be low enough to meet these conditions and also be high enough to drive T_3.

performance of basic circuit

It should be clear from the above that the d.c. levels which apply with no input connection depend on I_{b1}, I_{b2} and $(V_{be1} - V_{be2})$. Each of these gives its own contribution to the zero offset referred to the input, i.e. that voltage which, if applied as an input, would bring the output voltage to zero. Since the circuit is linear the separate effects may be added to give the correct overall figure, provided overload does not occur, i.e. no transistor cutting off or bottoming):

$+I_{b1}$ gives an apparent input of $-R_1 I_{b1}$
$+I_{b2}$ gives an apparent input of $+(R_4 \| R_5) I_{b2}$
$(V_{be1} - V_{be2})$ gives an apparent input of $+(V_{be1} - V_{be2})$

The total offset referred to the input is therefore

$$(R_4 \| R_5) I_{b2} - R_1 I_{b1} + V_{be1} - V_{be2}$$

offsets due to base current

To minimize the offset due to I_{b2} and I_{b1} each should be small, as assumed in the previous section. There is a lower limit to these values, set by circuit requirements and transistor β.

If I_{b1} and I_{b2} cannot be made negligible, it may be possible to match T_1 and T_2 to make $I_{b1} \approx I_{b2}$ and to design $R_1 = R_4 \| R_5$. This equalizes the base circuit resistance for T_1 and T_2 and the total offset from I_{b1} and I_{b2} cancels to give zero output with no input. This is a popular technique when using microcircuits but is not always satisfactory. One snag is that T_1 base does not rest at zero voltage; this may affect the signal source. Another snag is that a change of source resistance – often brought about by connecting and disconnecting the source – causes an output voltage change. An even worse snag in a circuit designed for a wide temperature range of operation is that the two base currents may change unequally even in a so-called matched pair when the temperature varies. It is always preferable, whenever possible, to make both $I_{b1}R_1$ and $I_{b2}R_4 \| R_5$ negligible when considered separately. Like most compensation techniques the matching idea is best used to make a marginal circuit fractionally better, not for the correction of gross differences.

V_{be} offset

Here the actual values of V_{be1} and V_{be2} cannot individually be reduced, being controlled by the threshold voltage for the semiconductor used – invariably silicon in this type of circuit. Matching must be resorted to if small offset is important, unless a setting-up potentiometer is permissible as described later. Fortunately V_{be} matching is successful and for tracking with temperature also can be good, a figure of 5 μV/degC for $(V_{eb1} - V_{be2})$ being obtainable from a dual transistor.

input impedance

The input impedance consists of two significant parts. Assuming infinite gain, T_2 base follows T_1 base identically so that T_1 and T_2 share the current provided by the emitter resistor R_3 in the same proportion whatever input is applied (within the linear range of circuit operation). When an input voltage e is applied, therefore, T_1 and T_2 base voltages both change by e volts and, provided $R_3 \gg r_e + r_b/\beta$, the emitter voltages also change by e. The current in R_3 must change by e/R_3 A, and the two base currents change by $e/2\beta R_3$ A if the emitter current is shared equally (usually so if V_{be} changes are intended to cancel). The input impedance is therefore $2\beta R_3$ if the internal amplifier gain is infinite.

When the assumption of infinite gain is not made there is an additional current change in T_1 base because the ratio in which T_1 and T_2 share the R_3 current changes, since T_2 base voltage is no longer identical to T_1 base voltage. If the internal gain is A, then the signal between bases, which is $V_{in} - V_{out}R_4/(R_4 + R_5)$, must be equal to V_{out}/A. This yields

$$V_{in} - V_{b2} = V_{in}\left(\frac{R_4 + R_5}{R_5 + (1 + A)R_4}\right)$$

which shows that the base-to-base signal is approximately equal to the input voltage divided by the loop gain, i.e. the ratio of internal gain to over-all gain. The input impedance due to this is therefore

$$\frac{2(r_b + \beta r_e)(1 + A)R_4}{R_4 + R_5}$$

(Since $(r_b + \beta r_e)$ or h_{11} is the input impedance of an earthed emitter amplifier, that figure has to be multiplied by twice the loop gain to give the correct result, the factor of 2 being appropriate for an emitter coupled pair (see Chapter 4 of EDH).)

In most practical designs the loop gain will be about 50 and the value of $(r_b + \beta r_e)$ will be in the region of 10 kΩ at the low operating currents used, say 0·1 mA, giving an input impedance of 1 MΩ from this source.

The input impedance also includes the base-collector circuit of T_1; this is normally several megohms (r_c) and can be ignored, but if a special attempt is made to increase the other components it could become significant.

When the $2\beta R_3$ contribution is noticeable it can be increased by replacing R_3 by a constant current source. Should this be necessary it is then doubtful whether this circuit has any advantage over an integrated circuit since its main attraction is cheapness and simplicity.

The above input impedance calculations refer to the circuit itself and do not include the base resistor, which appears directly in parallel. As indicated earlier, its value is determined by d.c. considerations because of the effect of I_{b1}.

In reality it is the parallel value of R_1 and any d.c. connected source resistance which has to satisfy the d.c. condition.

Therefore, if a source is permanently directly coupled to the input, R_1 may be omitted provided the source resistance is no larger than d.c. considerations permit. The input impedance is then not degraded by R_1. This cannot be done if the source resistance is sometimes high and at other times low since d.c. conditions will change (due to I_{b1}).

In the circuit of Fig. 13.1, low-frequency a.c. and d.c. performance is identical. Where only a.c. signals are of interest and the input impedance is noticeably affected by R_1 and R_3, a simple modification is often useful (Fig. 13.2). Here, C_A and C_B are bootstrap com-

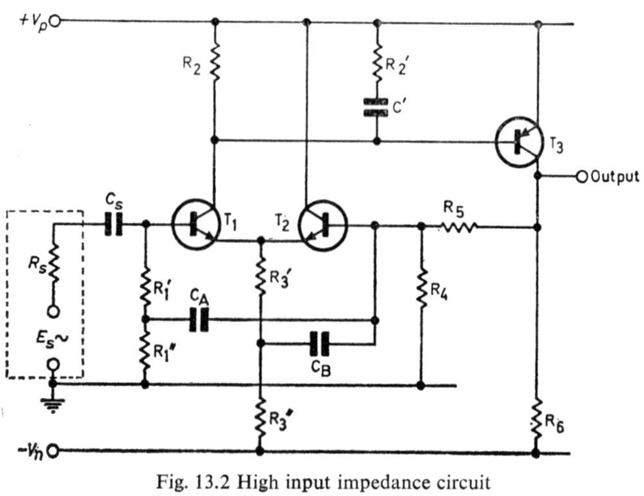

Fig. 13.2 High input impedance circuit

ponents and have the effect that equal signals are applied to each end of R_1' and R_3'. No signal current then flows in these resistors and they behave like much bigger resistors. This is a commonly used modification but care must be taken that a response peak does not occur. This is fully dealt with in Chapter 15 of EDH; it is sufficient to point out that the use of C_s, R_1', R_1', C_A produces a peak in over-all amplifier response of magnitude

$$\left[1 + \frac{C_A}{C_s} \cdot \frac{R_1' R_1''}{(R_1' + R_1'')^2}\right]^{\frac{1}{2}}$$

at a low frequency

$$f = \frac{1}{2\pi (C_A C_s R_1' R_2'')^{\frac{1}{2}}}$$

output impedance

Since the collector impedance r_c of a transistor is high, the output impedance for Fig. 13.1 is $(R_5 + R_4) \| R_6$ in the absence of feedback; with feedback as shown, this is divided by the loop gain giving

$$\frac{[(R_5 + R_4) \| R_6](R_5 + R_4)}{AR_4}$$

where

$$A = |\text{internal voltage gain}|$$

Drive capability is limited on positive-going outputs to the amount available from T_1 collector with the base current received from T_1. The limit is reached when all the current in R_3 passes through T_1, T_2 being just cut off; some is lost in R_2 and the rest drives T_3 giving

$$I_{b3} = \frac{V_{in} + V_n - V_{be1}}{R_3} - \frac{V_{eb3}}{R_2}$$

T_3 can drive βI_{b3}, but loses

$$\frac{V_{out}^+}{R_5 + R_4} + \frac{V_{out} + V_n}{R_6}$$

in driving R_4, R_5 and R_6. The available load current is therefore

$$\beta\left(\frac{V_{in} + V_n + V_{eb1}}{R_3} - \frac{V_{eb3}}{R_2}\right) - [V_{out\,max}^+/\{(R_5 + R_4) \| R_6\} + V_{in}/R_6]$$

giving

$$V_{\text{out}}^{+}\left(\frac{1}{R_2}+\frac{1}{R_6}+\frac{1}{R_5+R_4}\right)=\beta\left(\frac{V_{\text{n}}+V_{\text{in}}}{R_3}-\frac{V_{\text{be}}}{R_2}\right)$$

If this is a dangerously high limit that could damage T_3 or the load, a current limit can be set by inserting a resistor R_7 in T_3 collector lead; this limits T_3 collector current to $(V_{\text{p}} - V_{\text{out}})/R_7$ instead of $\beta I_{\text{b}3}$.

On negative-going outputs, the current is limited by R_6 which supplies the load and $(R_5 + R_4)$. The load current limit is therefore

$$\frac{V_{\text{n}} - V_{\text{out max}}^{-}}{R_6} - \frac{V_{\text{out max}}^{-}}{R_4 + R_5}$$

In some applications the output impedance is too high and the intended load causes severe loss of loop gain, degrading most of the performance parameters. Often, the drive capability is too small and can only be increased by decreasing R_3 which leads to either lower values of R_1, R_4 and R_5 or worse d.c. offsets. By the addition of an emitter follower T_4 (Fig. 13.3) a great improvement in both output impedance and drive capability can be obtained. For most practical cases the collector load of T_3 is now higher than previously so that loop gain is also increased. When it is desirable to limit the load

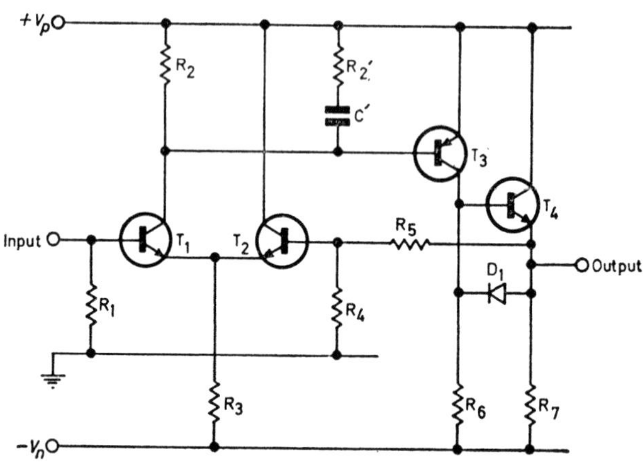

Fig. 13.3 Modified form of circuit of Fig. 13.1 for lower output impedance, higher gain and greater drive capability

current a resistor R_8 may be added in the collector lead of T_4. This limits the available current on positive swings to $(V_p - V_{out})/R_8$ and the negative swing current limit is determined by R_7.

hf stability

Owing to the lags introduced by the transistors and stray capacitances, the phase shift round the feedback loop may be sufficient to cause HF oscillation. This subject is covered in Chapter 8 of EDH, the practical cure being a *CR* network in parallel with R_2 which is made the dominant time constant. For audio applications, $R_2' \approx R_2/10$ and $C' = 470$ pF are adequate; where the ultimate in frequency response is required the values must be found by laboriously plotting the loop gain and phase, following the procedure described in the above reference.

gain variation and multi-stage operation

Because the d.c. output and input levels in a good design are nominally equal (and zero in the given circuit) direct coupling may be used between stages for either a.c. or d.c. use. The input resistor R_1 for all stages but the first sbould be omitted, its effective value with respect to the effects of I_{b1} being the output resistance of the driving stage.

If the d.c. version shown in Fig. 13.1 is used only for a.c. signals, direct coupling as suggested above has the disadvantage that small offsets and temperature drifts in the first stage become amplified in subsequent stages. This is often unimportant provided that when the a.c. signal is applied the later stages do not limit due to d.c. levels being too much in error. The d.c. offset present at the output can then be removed by final capacitance coupling.

When the build up of offsets is dangerous and would require unreasonable measures to cure, e.g. the use of a low-current high-β dual transistor in each stage, capacitance coupling to an input resistor may be used between each stage. A more economic and elegant solution is to add a capacitor in series with R_4; provided this is sufficiently large ($\omega_{min}CR_4 \gg 1$), the a.c. gain is unaffected but d.c. gain per stage is unity. The offset after n stages is then only n times that for one stage, rather than G^n where G is the stage gain (offset per stage and G are assumed identical in every stage).

effect of supply variation

Supply variation has no first order effect on the performance, provided ratings are not exceeded nor output swings limited.

Variations in V_p appear with similar magnitude at T_3 base and emitter and therefore cause no change in T_3 collector current.

Variations in V_n primarily cause T_1 and T_2 operating currents to change: T_2 current is held constant by the negative feedback loop at a fixed output so as to provide exactly the required drive for T_3, so only T_1 current changes, thus unbalancing T_1 and T_2. In a critical application there would be a discernible change in offset from this cause. The effect of R_6 (and R_7 in Fig. 13.3) is to compensate in part for this effect.

In practice, in non-critical applications supply variations of ± 50 per cent can often be tolerated; this is much better than most other three-transistor feedback circuits.

The circuit of either Fig. 13.1 or Fig. 13.3 is of most interest in d.c. and LF applications, with a useful over-all voltage gain from unity to a factor of about 30.

If R_5 is replaced by a load, the circuit becomes a voltage to current converter such that $I_L = V_{in}/R_4$; the load is 'floating' relative to the input and this limits the usefulness of the arrangement to meter, pen recorder and motor driving (see Chapter 15).

The circuit is not recommended for wide band applications in excess of 1 MHz owing to the difficulty of making accurate assessments of break frequencies. This means that loop stability has to be ensured by the use of a heavily dominant time constant (see page 157) which severely cuts the frequency band over which gain definition remains accurate. The procedure for calculating the required value, knowing the break frequencies involved is given in Chapter 8 of EDH.

Where wide-band response is important a similar circuit which, however, is easier to analyse at high frequencies is given in Fig. 13.4. Here T_3 is an emitter follower, the circuit having a better frequency response but less gain than before. Its susceptibility to

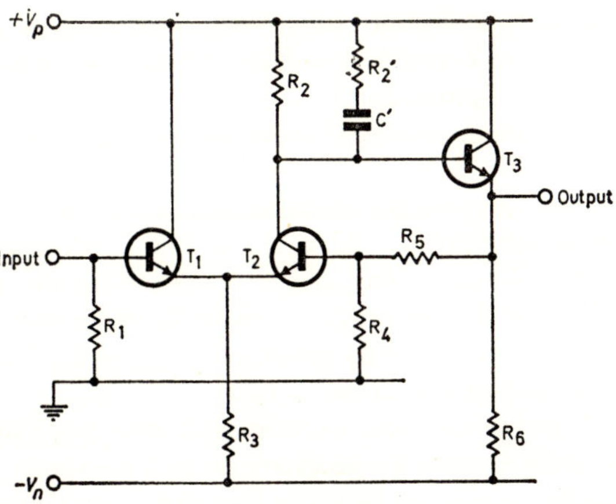

Fig. 13.4 Alternative circuit with lower loop gain but improved frequency response

variation in the positive supply is also greater, there being direct entry through R_2; this can be significant even though the effect is reduced by the gain of the loop.

protection circuitry

Since these circuits are often used alone, the source and load being outside the designer's control, it is advisable to include protection circuitry.

Input protection is achieved most simply by series resistance and shunt diodes (13.5) if the supply lines represent safe limits for T_1

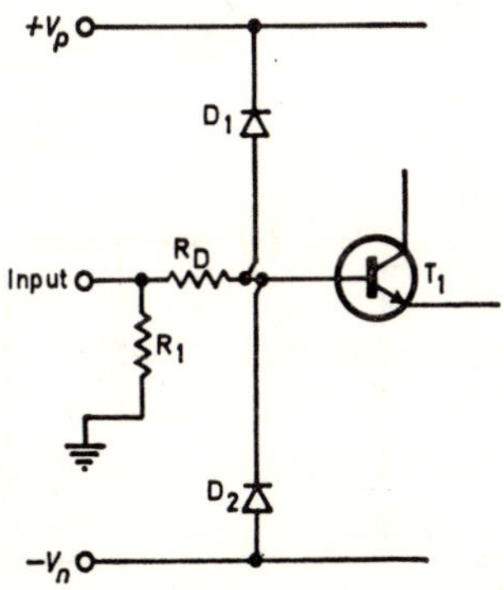

Fig. 13.5 Protection for excessive inputs

and T_2 reverse base-emitter rating. When this is not so, e.g. 2N930 transistors and 12 V supplies, additional base emitter shunt diodes may be added to T_1 and T_2.

Output protection against short-circuit load is readily provided by a resistor in T_3 collector (Fig. 13.1) or T_4 collector (Fig. 13.3), preventing excessive dissipation in the output transistor.

A further danger exists in the circuit of Fig. 13.3, if a short-circuit load or a heavily capacitive load is present when the signal goes negative. T_3 collector voltage falls, leaving T_4 either permanently or momentarily reverse-biased by possibly $V^{+}_{out\ max} - V_n$. This is cured by the shunt base-emitter diode in T_4.

In exceptional cases where the load can deliver large signals back to the amplifier it may be necessary to add resistance R_p and a catching diode to $-V_n$ as shown in Fig. 13.6. Note that the protection resistor R_p is within the feedback loop so that the output impedance under normal conditions is only slightly increased.

wiring

To obtain optimum results the following simple precautions should be taken. R_2 should be returned to V_p at the same point as T_3 emitter. The 'earth' connection of the source, R_1 if fitted, R_4, and the load earth return should meet at just one point. If this is not done there may be a signal between the earthy leads, giving gain errors (and in some layouts negative output resistance at d.c.).

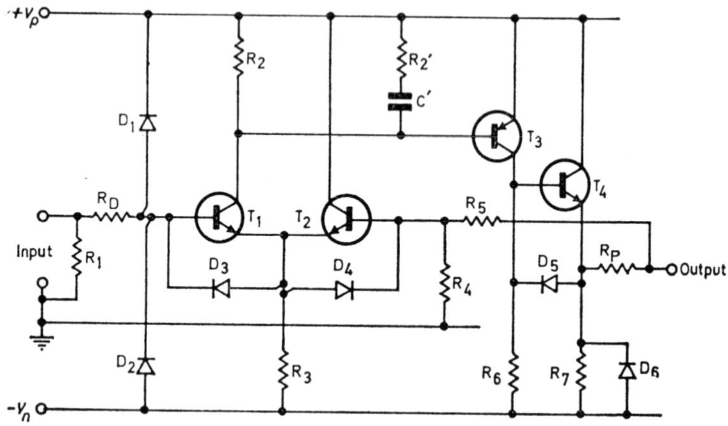

Fig. 13.6 Input and output protection

T_1 and T_2 should be mounted together and shielded from draughts if low drift is important. Even a plastic or fibre hood makes a very noticeable difference to drift performance in most environments.

detailed design

Design procedure is straightforward and should be clear already if the principle of operation is understood. The main point of difficulty is in deciding whether or not T_4 is required, i.e. whether the simple circuit of Fig. 13.1 or the more complicated form of Fig. 13.3 should be used. The quickest answer is to attempt a design using Fig. 13.1 and change to Fig. 13.3 only if really necessary, compromises having to be made between drift (or offset), output drive, loop gain and cost.

For once, design can begin at the input, the most critical case being a d.c. amplifier in which change of the d.c. level must be negligible with source connected or absent, and in which input impedance is to be high. This implies that the offset due to $I_{b1}R_1$ must be negligibly different from that due to $I_{b1}(R_1 \| R_s)$. At the same time R_1 is to be large (note that no bootstrapping technique can make R_1 appear to be large without also giving the appropriate offset; see Chapter 1), so that an immediate maximum value can be given to I_{b1}. The difference between V_{eb1} and V_{eb2} also contributes directly to offset voltage, and these considerations will normally suggest a transistor type and a suitable I_e, hence R_3. This I_e must drive T_3 which drives the load. At this point it becomes more obvious whether to add T_4, but this cannot finally be decided until the rest of the procedure is completed.

R_2 is chosen as usual from I_{CBO} considerations; its maximum value would be given by $R_2(I_{CBO2} + I_{CBO3}) = 0{\cdot}5$ V, where I_{CBO} is the value at the maximum junction temperature at which T_2 and T_3 are to operate. (For T_2 this usually is merely the maximum ambient temperature but T_3 may dissipate appreciable power.) A safety factor of at least 3 is advisable—it scarcely affects normal performance but allows an out-of-tolerance resistor or over-leaky transistor to operate. If the circuit is to operate above 10 kHz, R_2 may need to be smaller since it may represent, with stray capacitance, an appreciable time constant. This is calculated from $R_2(C_{cb2} + G_3C_{cb3})$ where G_3 is the voltage gain from T_3 base to collector, because the effective value of C_{cb3} is increased by this factor. The effect, if R_2 is too large, is that positive-going input and output signals are faithfully produced but signals with negative rates of change are distorted: $(C_{cb2} + G_3C_{cb2})$ can be charged by T_2 current increasing at a rate determined by R_3, but as T_2 tends to cut off, T_3 base rises on a time constant $R_2(C_{cb3} + G_3C_{cb3})$. $R_4\|R_5$ is chosen, like R_1, to give acceptable offset $I_{b2}R_4\|R_5$, and R_5/R_4 is designed to give the required gain $(R_5 + R_4)/R_4$. In Fig. 13.1, R_6 is decided as follows. T_3 is assumed cut off and with the load connected R_6 is designed to be small enough so that the load voltage is more negative than the required maximum negative output voltage. The margin should be generous if possible to avoid a tight tolerance requirement on R_6 which is quite unnecessary from other considerations.

Only now can the need for an extra transistor really be ascertained, by calculating the current required from T_3 to drive $R_5 + R_4$, the load, and R_6 all in parallel to the maximum required positive output voltage. If this current exceeds β_3I_{b3}, T_4 must be added. At this stage the loop gain should also be calculated to determine whether the required stability and accuracy of gain is adequate (an internal gain of about 400 is usually obtained). If the gain is insufficient, again the circuit of Fig. 13.3 with its extra transistor is indicated.

In using the Fig. 13.3 circuit the emitter components of T_4 are designed in the same way as the collector components of T_3 in Fig. 13.1. The extra resistor R_6, now the collector load for T_3 is calculated rather like R_2, i.e. if T_3 cuts off, T_4 must be cut off by this resistor against $(I_{CBO3} + I_{CBO4})$ tending to turn T_4 on. Again, speed considerations may indicate a smaller value since the maximum rate of fall obtainable at T_3 collector and hence T_4 base and emitter, is determined by $(C_{cb}T_3 + C_{cb}T_4)R_6$. If a faster output descent is required than this time constant would achieve, no amount of negative feedback will help—current is needed to drag this capacity negative enough in the time required and this current can only be supplied by R_6. R_7 is now determined by drive current considerations (as R_6 in Fig. 13.1).

Protection is now provided as required by the operating conditions to be used. As a general rule positive-going output current limiting by series resistance in T_4 collector (Fig. 13.3) or T_3 collector (Fig. 13.1) and a shunt base emitter diode on T_4 should always be included.

Even if not specified, short circuit load conditions are usually worth considering when protection is simple.

practical example

To illustrate the above procedure, the following design example will be calculated.

Source resistance: 100 Ω or ∞, i.e. source may be removed.
Source voltage ±1 V d.c.
Amplifier Performance—30 mV referred to input, with source resistance either 100 Ω or ∞

Temperature drift:	<1 mV/degC
Input impedance:	>5 kΩ
Gain:	5 ± 5 per cent
Supplies:	±20 V ± 10 per cent
Load:	⩾2 kΩ

Following the procedure detailed above, it can be assumed that R_1 must exceed 5 kΩ and could tentatively be given the value 6·8 kΩ. This assumes, with some pre-knowledge of the normal properties of this type of circuit, that the extra components of input resistance will be high enough to maintain the required 5 kΩ. Clearly R_1 cannot be omitted since the source is at times disconnected.

To meet the offset of 30 mV, $I_{b1}R_1 < 30$ mV, i.e. $I_{b1} < 4\ \mu$A. This value of I_{b1} would allow no V_{be} differential and a figure of 1 μA, with $V_{be1} - V_{be2}$ less than 10 mV is more realistic. Since the initial V_{be} tolerance of separate transistors is greater than 10 mV either a dual type, e.g. 2N3680, must be used, or a 'set zero' potentiometer added to the circuit. The temperature drift required would be met by either arrangement because each V_{be} separately changes only by about 2·5 mV/degC and I_{b1} will change about 2 per cent/degC.

Assuming at present that a dual transistor such as the 2N3680 is used, its β at 1 μA base current is greater than or equal to 100. The total emitter current should then be at most 100 μA: balance between the two emitter currents cannot be guaranteed at all inputs so it is safest to assume it may almost all flow into T_1 (if balance were certain the total current could be 200 μA). Therefore, $R_3 \geqslant V_{n\,max}/100 \times 10^{-6}$, i.e. greater than or equal to 220 kΩ, e.g.

270 kΩ. Several safety factors have been used to ensure that the offset specification is met. If difficulties follow in other aspects of the design these can be reconsidered using less generous margins.

R_2 is now designed assuming reasonable I_{CBO} figures, e.g. 1 μA at 50°C, giving $R_2 < 0{\cdot}5/2 \times 10^{-6}$, i.e. 270 kΩ, e.g. 82 kΩ, again a very large safety factor. With a normal V_{be} for T_3 of 0·7 V, R_2 will carry 9 μA so that the available current to drive T_3 is at worst:

$$\frac{V_{n\,min}}{R_3} - 9\mu A = \frac{18}{270}.10^3 - 9 = 57\ \mu A$$

This assumes that T_1 is allowed to reach cut off so that T_2 can use all the current in R_3 to drive T_3. A much more desirable condition would leave T_1 with at least 20 μA, leaving 37 μA to drive T_3 (because T_1 β will fall at currents below 20 μA).

The feedback resistors should not exceed about 5 kΩ in parallel resistance in order not to degrade input offset—up to this value they tend to improve the offset but this is unreliable because the emitter currents are not always balanced, giving unequal base currents even if β is matched. This gives possible values of 5·1 kΩ‖240 kΩ for R_4, 20 kΩ for R_5, to give close to 5 for nominal gain.

Using Fig. 13.1, with a maximum negative output of −5 V and a $V_{n\,min}$ of 18 V, the value of R_6 is given by

$$\frac{13}{R_6} \geqslant \frac{5}{R_{L\,min}} + \frac{5}{(R_4 + R_5)}$$

i.e.

$$13/R_6 \geqslant (2{\cdot}5 + 0{\cdot}2)\ \text{mA}$$

or

$$R_6 \leqslant 4{\cdot}7\ \text{k}\Omega, \quad \text{e.g. } 3{\cdot}9\ \text{k}\Omega$$

When maximum positive output occurs the current to be provided by T_3 is at worst

$$\frac{5}{(R_4 + R_5)} + \frac{5}{R_{L\,min}} + \frac{(5 + V_{n\,max})}{R_6}$$

i.e.

$$(0{\cdot}2 + 2{\cdot}5 + 6{\cdot}9)\ \text{mA} = 9{\cdot}6\ \text{mA}$$

The β of T_3 is therefore required to be (9·6/37)1,000, i.e. about 250.

It is clear that a special transistor would have to be found to meet this β specification. The decision can now be made to add T_4 as in Fig. 13.3; alternatively the rather generous tolerances can be pared down by more accurate assessments of drift in the hope that T_4 may be omitted. In the present example it is safer to add the extra transistor.

This gives R_7 in Fig. 13.3 the value 3·9 kΩ as its function is similar to R_6 in the simpler circuit. The new R_6, like R_2, could be in the limit 270 kΩ and again 82 kΩ gives a large margin.

A protection resistor in the collector of T_4 must limit the current to a safe value in the event of a short-circuit load. Assuming T_4 bottoms and that it can withstand 50 mA without damage, then the resistor can be $(V_{p\ max}/50)$ kΩ = 440, e.g. 470 Ω.

Choice of transistors is now easy since $\beta_3\beta_4$ needs only to be about 250 (like β_3 in the three-transistor circuit) and almost any silicon *p-n-p* type of adequate rating will suffice for T_3, e.g. 2N3702, 2N2906A, etc.); T_4 may be BC108, 2N2369, etc.).

zero offset

In a practical design the question of zero drift often allows two alternative designs. One uses a matched pair for T_1 and T_2; the other uses an unmatched pair and then usually requires a set-zero potentiometer arrangement, because initial V_{be} unbalance is too great although temperature tracking may be satisfactory. The decision is not always easy to make, since the higher component cost of the matched pair is partly offset by the cost of setting up and the degraded reliability due to the potentiometer.

Generally, in a commercial application where all specifications are met by unmatched devices, except initial setting up, a zero-setting arrangement is economic. In industrial and military applications the extra reliability and better performance in other respects also (temperature drift) favours the matched pair.

Suitable zero setting circuits are described in Chapter 15 in connection with a similar circuit.

CHAPTER 14

inverters

THE title refers to circuits in which a d.c. supply is converted into a.c. by electronic switching. Various techniques are used and each has advantages in particular applications. For power below 50 W, and where efficiency is unimportant but output voltage is required to be well defined, the self-oscillating form given in Fig. 14.1 is commonly used.

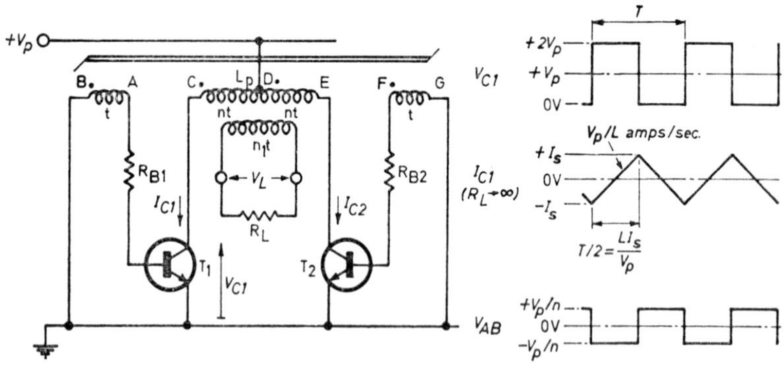

Fig. 14.1 Basic inverter

This basic form has several variants with only slightly different properties; they have been described by Wolfendale, Baxandall and others and are compared below. Another lesser known form, described in the final section, is often useful when the supply voltage is high.

standard circuit

Figure 14.1 operates as follows. Assuming T_1 conducts momentarily, T_1 collector voltage falls causing the base drive voltage

V_{AB} to rise thus further increasing T_1 conduction until T_1 bottoms provided the base drive is sufficient.

Meanwhile T_2 base is driven negative and T_2 collector to $+2V_p$ by transformer action. Since the voltage across CD is V_p, A is now at $+V_p/n$, F at $-V_p/n$ and V_L is $V_p(n_1/n)$. Bottoming will be maintained provided

$$\beta_1\left(\frac{V_p}{n} - V_{be}\right)\Big/R_{\beta_1} > I_{c1}$$

where I_{c1} is the current in T_1 collector due to R_L, i.e. $V_p n_1^2/(n^2 R_L)$.

Under these conditions the inductance between C and D, namely L, is being driven by a constant voltage V_p and its current rises (if L is constant and loss-free) linearly at a rate

$$\frac{di}{dt} = \frac{V_p}{L}$$

This current causes I_{c1} to increase steadily, resulting in two alternative modes of circuit behaviour.

If L is a near-linear inductor, and in particular does not reach magnetic saturation, then I_{c1} continues to increase until the base drive, which is constant at $[(V_p/n) - V_{be})]/R_{B1}$, is incapable of maintaining bottoming. When this occurs V_{CD} decreases thus decreasing V_{AB} which determines the base drive. This decrease in base drive allows V_{CD} to reduce further and simultaneously the reverse drive on T_2 base decreases. The sequence continues until T_2 turns on and T_1 is driven off, the transition being rapid because of the positive feedback action just described.

T_2 now bottoms and a similar action follows. The time for which each transistor conducts is determined by its β since this decides how large the current in the collector inductance has become before the base drive is inadequate.

This mode of operation therefore has a mark/space equal to the ratio of the two values of β.

If on the other hand L is designed to reach saturation before base drive fails, the collector current still rises initially at a rate

$$\frac{di}{dt} = \frac{V_p}{L}$$

When magnetic saturation is reached the effective value of L is much reduced (to the value obtained if the core is removed) and a very sharp rise of current follows. Whatever the value of β, the base drive is unable to maintain bottoming and T_1 turns off, turning T_2 on.

Under these conditions a symmetrical mark/space is obtained provided the transformer core has no standing magnetic bias, saturating at the same ampere-turns in each direction.

The second operating condition is clearly preferable since a β-dependent mark/space has many unwanted side effects. For instance, if the output is to drive a load which operates on the fundamental component only, this component varies with mark/space, so the effective output voltage is badly defined.

The design procedure is straightforward and well described by Wolfendale. In practice the equations used (and quoted above) are inaccurate because the magnetic circuit is non-linear, giving a cubic curve for i against t. This causes the predicted frequency to be in error by as much as two to one.

starting circuits

The basic circuit is not self-starting since both transistors are initially cut off. As in other circuits which may not start, the cure can consist of a starting pulse which turns one transistor on momentarily. It is a satisfactory method only if a subsequent return to its static condition is not disastrous and if recovery can wait until an operator restarts it.

This can be achieved by a capacitor which provides a positive pulse to one transistor base. The pulse must disappear before that transistor is required to turn off again, or it must be small enough to be overridden by the reverse base drive.

An alternative method is to bias one or both transistors slightly 'on' by resistors between V_p and the bases. These must be chosen with care: at one extreme of tolerance there is a danger that the transistor is not quite conducting, and at the other, the subsequent reverse drive may not exceed the forward bias to turn the transistor off. There is usually a suitable value except when V_p is highly variable.

Another snag is the constant, wasteful power drain through the extra resistors. This can be overcome by the use of a diode-transistor network which effectively disconnects the resistor when operation is well under way (Fig. 14.2). This method is expensive in components but is compensated for by the fact that the diodes help to damp collector voltage transients. At any time when oscillation is not

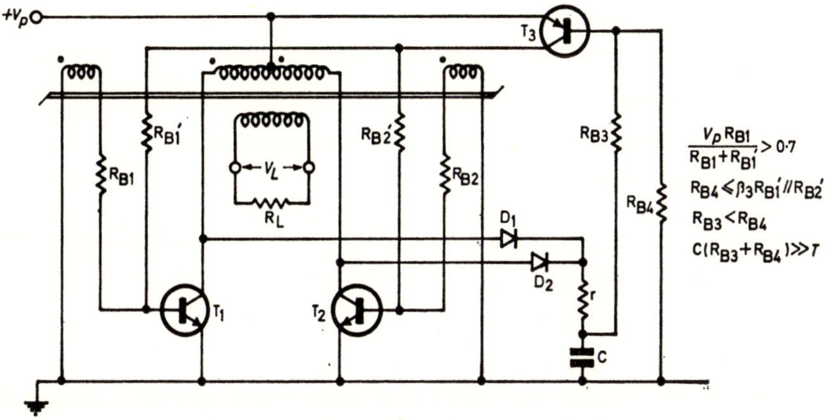

Fig. 14.2 Self-starting circuit (low power loss)

present, T_3 is bottomed by R_{B4}. Resistors R'_{B1} and R'_{B2} are designed to bring T_1 and T_2 into conduction, i.e.

$$V_p \frac{R_{B1}}{R_{B1} + R'_{B1}} > 0{\cdot}7 \text{ V}$$

When oscillation begins, C charges to $+2V_p$ (slightly reduced by current limiter r) and T_3 is reverse biased by ensuring $R_{B3} < R_{B4}$. Since

$$R_{B4} \approx \beta_3(R'_{B1} \| R'_{B2})$$

the power loss of the starting circuit is improved by β_4.

efficiency

Neither this circuit nor its derivatives described below are particularly efficient in operation. This is partly due to the power normally lost in the starting circuit. The main power loss, however, is caused by the saturation current required by the transformer (the fact that it is 'reactive' is no consolation to the transistors and power supply which have to supply it!); attempts to keep this small result in difficult windings of many turns, giving high inductance, low operating frequency and large volume and weight. Some power is also lost in R_{B1} and R_{B2}. If this is minimized by choosing a high value of n, small driving voltage and small resistors, the base drive current is ill-defined, being proportional to $[(V_p/n) - V_{be}]$.

For high efficiency, these circuits are therefore not recommended, a separate low power driver coupled to a power output stage being more suitable.

The circuits described here have the merit of simplicity with efficiency of about 60 per cent.

winding arrangements

Several versions of Fig. 14.1 have been devised for particular requirements. Figure 14.3 illustrates a design due to Baxandall in which the base drive is obtained from one continuous winding which

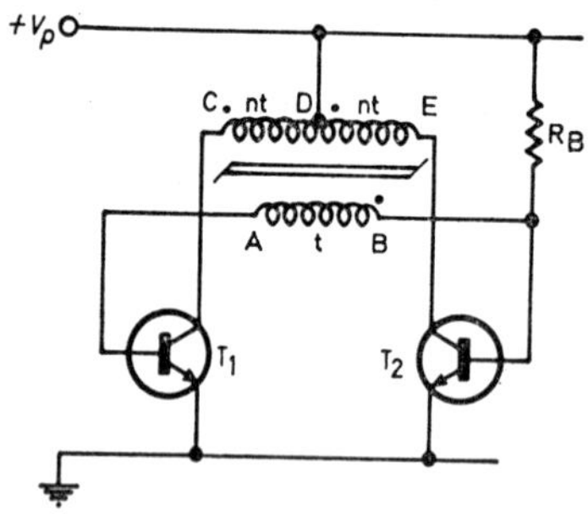

Fig. 14.3 Self-starting form of circuit of Fig. 14.1

serves both transistors. The basic operation is unchanged but when A becomes positive with respect to B, A is caught at about 0·7 V by V_{BE1}, forcing B, i.e. T_2 base, to reach $-V_p/n$. At this time T_2 is cut off and the current in R_B, i.e. $[V_p + (V_p/n) - V_{be}]/R_B$ must flow through winding AB into T_1 base. When B becomes positive with respect to A on the other half cycle, T_1 is cut off and R_B current $(V_p - V_{be})/R_B$ flows.

In this design the turns ratio n may be very large since only about 1 V across AB is needed to ensure switching. This voltage is not used, as in Fig. 14.1, to define the base drive but merely to cause the opposite transistor to turn off.

As drawn, the winding passes the R_B current only when T_1 conducts and passes only leakage current when T_2 conducts. The core is therefore subjected to a steady as well as an alternating field and this slightly affects the collector currents for saturation and hence mark/space. In most practical applications this is insignificant; if required, winding AB may be centre-tapped as before and R_B connected to it instead of to A or B.

The advantages of this configuration are that starting is certain due to the strong forward bias of both transistors; base drive is very well defined if $V_p \gg V_{be}$; the base winding has fewer turns and the centre tap may often be omitted. On the other hand, power loss in R_B is much increased.

Another modification is based on the idea of joining the collectors, rather than the emitters of T_1 and T_2, so that a common heat sink may be used without the drop in thermal conductivity which results from the use of insulating washers.

Referring to Fig. 14.1 it is evident that this can be done provided the collector windings are transferred to the emitter circuit while keeping the base windings returned to the individual emitters.

The base drive then turns on collector (and emitter) current and since the same winding CD is in series with the transistor and the supply as before, the transistor operation is unchanged (Fig. 14.4). It now becomes obvious that the connections and phases of the windings are equivalent to a continuous winding.

This gives a simple tapped winding (or two tapped windings each comprising a base winding and a collector winding) as well as the possibility of a common heat sink for T_1 and T_2, without insulating washers. Ideally a negative supply would be used so that the heat sink can be earthed. Design is of course identical to that of the original circuit.

When a sine-wave output is required it is not sufficient merely to

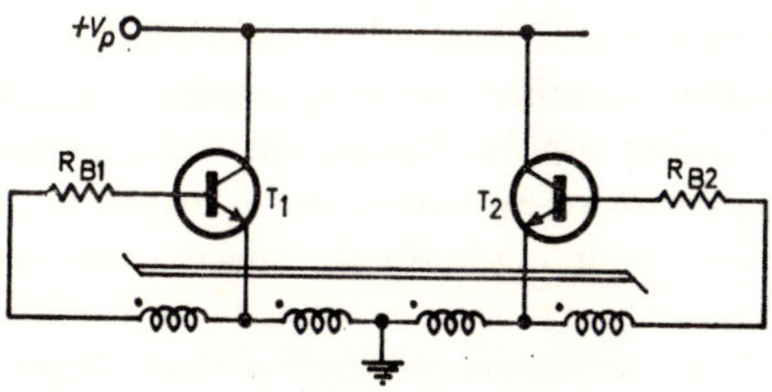

Fig. 14.4 Variation of basic inverter

tune the collector winding by capacitance. This would place a parallel capacitance load on the collector circuit which would either prevent bottoming or cause very large currents to flow near the start of each half-cycle. Baxandall has described an ingenious solution to this problem by inserting a large choke in series with the supply. Now each transistor can bottom in turn with no current surge, resulting in the waveforms shown in Fig. 14.5.

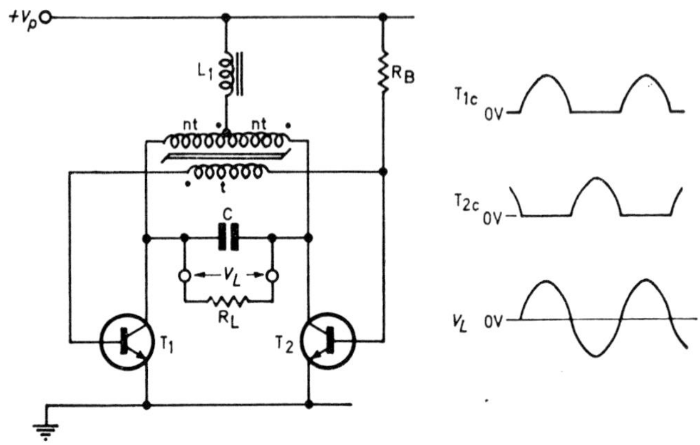

Fig. 14.5 Inverter for sine-wave output

This enables a reasonably sinusoidal output to be obtained with the efficiency of a normal saturating inverter. Extra constraints are placed on the transformer which must have small enough inductance to give good Q in the presence of the load.

The series inductor L_1 can be difficult to design because it carries the direct current of the collector circuits and is in danger of saturation.

The design procedure is fully described in Baxandall's excellent paper and will not be repeated here.

A problem that is encountered occasionally is the need for an inverter driven from a rather high supply voltage, e.g. 300 V, possibly derived from a mains supply. The circuits so far considered have the snag that the collector of each transistor is driven to $+2V_p$ when cut off. Moreover, the coupling between collector windings is imperfect and the leakage inductance, combined with stray capacitance, causes the voltage waveform to reach a much higher voltage as it 'rings'. In the application mentioned above, the collector voltage rating should be at least 1 kV, an unreasonable requirement with presently available transistors.

In such cases the series arrangement in Fig. 14.6 has two advantages. Firstly the transistor collector rating, ignoring overshoots due to leakage inductance, is only V_p, not $2V_p$. Secondly the over-

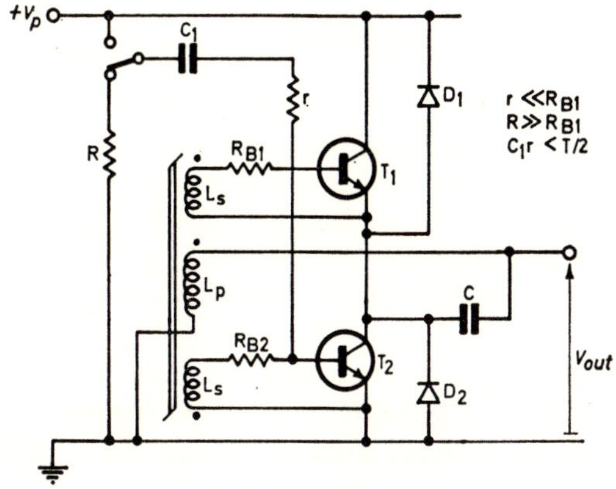

Fig. 14.6 Series inverter

shoots due to leakage inductance are readily clipped by the diodes shown—this is not possible in the preceding circuits as there is no available clipping voltage.

This circuit is not self-starting and, as before, forward bias or a starting switch can be added.

detailed design

Although the principle of design is the same as for the conventional push-pull inverter which is well described in the articles referred to, a design example for this series form may be useful.

The design requirement may be imposed in two ways. For a d.c.–d.c. inverter, achieved by bridge rectifying the a.c. output, the frequency is often left to the designers discretion. In general, a higher frequency leads to a smaller magnetic core but in the limit causes difficulty in transistor power dissipation during switching (which is a larger proportion of the cycle) and in rectifying (due to diode hole storage). For a d.c.–a.c. inverter the frequency is specified and this, with the output power requirement, determines the core size. In this case, owing to a limited choice of core materials and standard core sizes, the design process is iterative, three attempts usually being required.

Suppose $V_p = 40$ V, and $V_{out} = 18$ V d.c. into a load of 36 Ω. The series circuit can be used with advantage since a transistor of 50 V V_{ce} rating may be used, the normal circuit requiring at least 80 V (in practice 120 V is more realistic).

In Fig. 14.6, if $V_p = 40$ V, then V_{out} (a.c.) = 20 V pk, and a bridge rectifier across L_p will deliver about 18 V d.c. as required. In passing, it should be remarked how easy it is to convince oneself (and others!) that a 40 V pk-pk square wave across a winding such as L_p should give 40 V d.c. when rectified in a bridge. Such an illusion is readily dispelled by regarding one end of the winding as fixed to earth; the other end then clearly moves alternately to +20 V and −20 V, each time forcing the reservoir capacitor to 20 V.

In this example, the series circuit therefore has the further advantage that no extra transformer windings are required to supply correct voltage to the load. However, neither load terminal may be joined to an input terminal since this would short circuit part of the rectifier bridge. If complete isolation between input and load is required, or if there must be a specified common terminal, another winding must be provided.

The first design step is to calculate L_p, having made a guess at a suitable operating frequency. Bearing in mind the earlier comments on high frequency operation, a figure of about 10 kHz is suitable: rectifiers are reasonably efficient at this speed and switching times even as long as a few microseconds are but a small proportion of a period.

The frequency is related to circuit values as follows. The voltage across L_p on each half cycle is $V_p/2$ and the resulting rate of change of current in L_p is, from '$E = L(\mathrm{d}i/\mathrm{d}t)$',

$$\frac{\mathrm{d}i_{L_p}}{\mathrm{d}t} = \frac{V_p}{2L_p}$$

The time taken to complete a half cycle ($T/2$) is the time required to change the current in L_p from saturation current in one direction to saturation current in the other direction. This gives

$$\tfrac{1}{2}T = 2I_s\frac{2L_p}{V_p}$$

(see Fig. 14.7), i.e.

$$T = 8L_pI_s/V_p$$

With the given values, $T = 100\ \mu$sec, $V_p = 40$ V, so that

$$L_pI_s = TV_p/8 = 0{\cdot}5 \times 10^{-3}$$

It is now possible to select a core known to be of the correct order for this type of application and let the core determine the value of I_s, having made $L_pI_s = 0{\cdot}5 \times 10^3$. Alternatively the maximum tolerable I_s can be decided and the core specified.

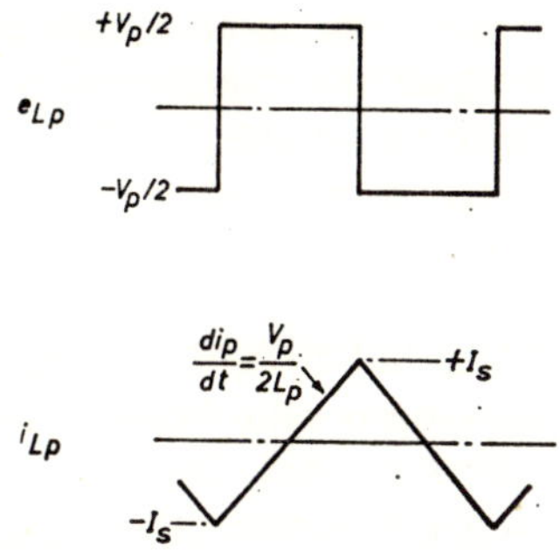

Fig. 14.7 Inductor waveforms for circuit of Fig. 14.6

The more useful approach is usually a combination of the two – a core is 'tried' in theory and if I_s exceeds a reasonable figure a different core is selected, the first try having indicated the correct direction to take.

In deciding I_s, it is clear that the current required for saturation is wasteful; the transistors have to drive it in addition to the normal load, thus needing extra base drive, and V_p has to supply all the extra current. It may, however, be impossible to achieve small I_s on a small core because of the difficulty of accommodating the large number of turns required without resorting to light wire gauge, giving high resistance. A reasonable criterion is to make I_s about 10 per cent of the full load current.

Taking

$$I_s = \frac{V_{out}\,(\text{d.c.})}{10R_L} = 50\ \text{mA}$$

gives

$$L_p = \frac{1}{100}\ \text{H} = 10\ \text{mH}$$

These figures are reasonably compatible with a small ferrite toroid—recommended for these circuits because of their low price—in which a typical core has L = 30 turns for 1 mH [and $L \propto$ (turns)^2] and about 5 ampere-turns for saturation.

Here 90 turns would be used, giving I_s = 45 mA, L_p = 10 mH. A suitable core is the MM454, measuring about 1″ O.D.

The transistors must have $V_{ce} \geqslant 40$ V, useful β and low $V_{ce\,(sat)}$ at $I_c = 0{\cdot}55$ A.

With the bridge rectifier used here, all the load current is supplied alternately by T_1 and T_2. Each transistor carries a mean current of only 0·275 A and this is also the mean supply current as this flows only through T_1. However the characteristics of the transistors when conducting have to be suitable for the current then flowing, i.e. 0·55 A. The 2N2221A is adequate with β of 25 min at 0·5 A, $V_{ce(sat)} = 1$ V max and $V_{be(sat)} = 2$ V max at $I_b = 50$ mA, $I_c = 0{\cdot}5$A.

Since V_{be} may be 2 V at full drive, the voltage from L_s should be at least 3 V pk, giving a turns ratio between the output and base windings of less than 20 : 3, e.g. 6 : 1. This gives an actual peak drive of $(V_p - 2V_{ce\,(sat)})/6 = 3$ V of which $(3 - V_{be\,(sat)}) = 1$ V is useful on a limit transistor. R_{B1} and R_{B2} are now given by $R_{B1} = R_{B2} = 1/I_B = 20\ \Omega$. The coupling capacitor C must couple the square wave given by T_1 and T_2 to the load, which effectively appears across L_p ,without excessive discharge during each half period. Only a surprisingly small amount of droop is permissible from this cause because of the resulting ripple and loss of amplitude in the d.c. output, and $CR_L \geqslant 100T$ is usually required. This gives $C \geqslant 100 \times 100 \times 10^{-6}/36 \approx 300\ \mu$F, *rated at 0·275 A ripple current*, and at least 20 V working.

For the application which led to the quoted example a starting switch was deemed satisfactory rather than a forward biasing arrangement. This consisted simply of the starting capacitor and resistor shown; note the recharge path through the resting contact, without which a perfect capacitor might start the circuit only once (for ever!) since its charge thus acquired would remain stored.

performance

It must be emphasized that designs of inverters using magnetic saturation are imprecise in prediction of operating frequency. The idealized waveforms representing the rate of current build up are not met in practice due to series effective resistance in the winding and non-linearity in the magnetic circuit long before saturation is reached. When saturation is reached, the current then rises more rapidly but not at the infinite rate usually assumed since the inductance of the winding in air is still present.

Added to this are the normal tolerance in permeability which leads to uncertainty in L_p and the much larger spread in the ampere turns at which saturation occurs.

The result is that the frequency of operation and the value of I_s are likely to be in error by as much as 100 per cent. This is usually unimportant in a d.c.–d.c. inverter provided the $V_{ce\,(sat)}$ voltages are insignificant. Output voltage and impedance are then dependent only on winding losses and the rectifiers as in a mains-derived supply.

In the series inverter used as a design example, efficiency is best at full load since some losses are fixed. The losses include 45 mA for I_s (1·8 W), 1 V each for V_{ce} (0·55 W), 2 V each for V_{be} (0·1 W), 1 V each diode in the bridge (1 W), totalling 3·45 W. Load power is 10 W, giving an efficiency of about 74 per cent.

If high efficiency or accurate frequency is required then a separate oscillator and output stage should be used.

protection

Transistor failure is unfortunately common in inverter circuits unless special protection measures are taken. There are two main failure mechanisms: excessive V_{ce} immediately after cut off, provoking avalanche action, and excessive base emitter voltage due to switching spikes.

The first often occurs because the transistor $V_{ce(sus)}$ rating is not quoted or has been ignored by the designer. If this voltage is exceeded even for a short time, a large current flows in the collector circuit and does not revert to its normal level when the excessive voltage is reduced below $V_{ce(sus)}$. Destruction of the transistor follows. In inverters it is essential to use only those transistors with a specified acceptable value for $V_{ce(sus)}$.

The second is caused by the very fast transition rates which occur in inverters due to the large amount of positive feedback. These fast transitions in circuits containing stray capacitance, inductance and leakage inductance, cause large oscillatory voltages to appear across the base windings. It is possible to clip these, using a fast diode back-to-back with the base-emitter junction of each transistor. It must then be realized that the base winding must now supply twice the normal current. Diodes cannot be placed directly across the windings since the diode for T_1 base would clip the drive required to forward-bias T_2 and vice versa, through the coupling of the base windings. Resistors and capacitors across the base windings can sometimes be added with advantage but calculation of their value is impractical since the leakage inductance, strays and Q of the spurious resonance are unknown.

Although fast transistors have the advantage that only small power loss takes place because of fast transitions, they aggravate the spike problem. It is safest therefore to use transistors which are no faster than necessary.

interference

Inverters are notorious in causing interference to neighbouring equipment. This is partly due to the high-speed switching transients which radiate readily and partly due to magnetic field leakage at the operating frequency. Acoustic interference is also a feature of many inverters due to core vibration.

Radiated interference from the high-frequency components can be reduced by electrostatic screening using if necessary two enclosures of thin material in preference to one thicker screen.

Magnetic leakage is minimized by using a toroidal or pot core for the transformer. Ferrite toroids have the advantage of cheapness and strip wound toroids of magnetic alloy are much lower in price than the equivalent lamination stampings. For medium quantities, say 20, the cost of winding toroids offsets the saving in hardwear. Small quantities are usually wound by hand and larger quantities are worth the setting up time required by a toroidal winding machine. Most inverter transformers up to a few watts rating can be wound readily by hand for laboratory experiments using small ferrite toroids, unless low-frequency operation (<1 kHz) is obligatory.

CHAPTER 15

meter-driving circuits

THE driving of an indicating meter or direct-acting pen recorder is a very common requirement in electronic equipment. Problems arise which are different from those encountered in normal circuit design; these can, however, readily be overcome by a simple feedback technique which has become the standard arrangement for this purpose. There still remain several design points that require special attention and these are dealt with in the following sections.

meter characteristics

Except in special cases, the ideal meter indication is linear with respect to the voltage or current to be measured. This restricts the designer generally to the use of a moving-coil instrument, the important characteristics of which are listed below.

1 The movement is current operated; this means that when the winding resistance changes due to temperature, the calibration remains approximately true in terms of coil current rather than coil voltage.

2 Winding resistance, and consequently the voltage required for full-scale deflection, is rarely specified to better than several per cent and in any case has a large temperature coefficient.

3 Calibration varies according to whether or not the meter is mounted on a metal panel by an amount, dependent on the magnet design, which can be as high as 20 per cent. Panel thickness and the closeness of fit between meter and panel can also cause a few per cent variation in sensitivity.

4 There are many other hidden specification points to be considered when choosing a meter, such as linearity, accuracy at less than half full-scale, temperature coefficient, response time, ageing, effect of mounting position and vibration. The details can only be found by careful reference to manufacturers' data; suffice it to say that to obtain a true reading to within one per cent of the applied current is exceedingly difficult when these factors are examined.

5 Meter response time or damping factor is greatly influenced by its sensitivity, physical size and driving impedance. This is obvious on realizing that the meter movement is really a motor. Other things being equal, a motor which takes high input power and has low inertia will change its armature position faster than one of low power and high inertia. This shows up clearly on comparing a 1-inch 10 mA meter with a 3-inch scale meter of less than 100 μA f.s.d. The effect of driving impedance is seen most vividly when applying a short-circuit across the input terminals of a motor which has been driven to high speed and allowed to free run with the source removed. This so-called dynamic braking slows the motor greatly. A similar if less dramatic demonstration can be given by holding a small moving coil meter, rotating it sharply through about 45° and observing the antics of the pointer with an open- and then a short-circuit across the meter terminals. For fast response, the driving impedance should be very high compared with the coil resistance.

6 A.C. METERS Apart from non-linear movements, as used in moving-iron and hot-wire meters, an a.c. meter normally consists of a rectifier bridge built into the case of a moving coil meter. Since many industrial users are more concerned with buying an ammeter with low voltage drop, rather than good RF performance or good accuracy at high temperatures, the rectifiers used may be selenium, copper oxide or germanium. These rectifier meters are rarely suitable for use in an electronic equipment because of the slow response (except germanium) and high reverse leakage giving poor accuracy. For a.c. measurements in electronic equipment it is therefore normal practice to use a d.c. moving coil movement and to rectify by electronic circuits as described below.

7 OVERLOAD Most moving coil movements will withstand a current equivalent to 3 × f.s.d. without damage or change of cali-

bration. Above this level the pointer and its mounting are usually damaged if a positive input is applied with the pointer initially near zero; the coil itself will generally accept a 10 × f.s.d. current.

It is important in the design of meter drives that switch-on and switch-off transients are restricted to 3 × f.s.d., although inputs known to be negative deflection from zero may be as high as 10 × f.s.d. This is only a general guide and the manufacturer must be consulted for accurate figures.

special problems caused by meter characteristics

The peculiarities of moving coil meters of particular interest to the circuit designer are items 1, 2, 5 and 6 above. They all point to the desirability of driving from a high impedance circuit, which is equivalent to driving the meter with a known current which does not change when the meter resistance changes.

This renders the reading unaffected by resistance changes of the winding due to temperature and to initial tolerance between one meter movement and another. It gives the fastest response of which the movement is capable (it is always easy to slow the response of the driving circuit electrically if required). In a.c. applications it enables a standard d.c. movement to be used with external silicon rectifiers, having good RF performance and low leakage, since with defined current drive the large voltage drop of such rectifiers gives no reading error.

A good test for a meter drive circuit is, therefore, that its reading with a fixed input should change negligibly if extra resistance equal to the coil resistance is added directly in series with the meter. It is also necessary of course that the reading has the correct value at all specified input levels, i.e. the current definition is accurate to the desired tolerance.

a.c. operation

Alternating currents and voltages are often defined in terms of their r.m.s. value. With pure sinusoidal waveforms this is directly related to the average and peak values by simple constants:

$$I_{\mathrm{r.m.s}} = \frac{\pi}{2\sqrt{2}} I_{\mathrm{av}} = \frac{I_{\mathrm{pk}}}{\sqrt{2}}; \qquad v_{\mathrm{r.m.s}} = \frac{\pi}{2\sqrt{2}} v_{\mathrm{av}} = \frac{v_{\mathrm{pk}}}{\sqrt{2}}$$

where I_{av} and v_{av} are the magnitudes of the average current and voltage over a complete positive or negative half-cycle.

In making an electronic voltmeter or ammeter there are considerable difficulties in measuring r.m.s. values directly and presenting the answer as a linear indication on a moving coil meter. It is normal practice to measure the average value of the signal waveform and to use the appropriate scale factor so that the meter is scaled in terms of r.m.s. This implies that readings of non-sinusoidal waveforms are incorrect by the ratio of the form factors of the sine-wave and the input waveform, because the average value is being measured.

When it is important that true r.m.s. be measured, the circuit design is much more difficult and beyond the scope of the present discussion. Several different approaches are in common use; perhaps the most usual makes use of an indirectly heated thermistor, heated by an amplified form of the input signal. The thermistor output is then a measure of the heating effect, i.e. r.m.s. value of, the input signal. A second thermistor is used to compensate for ambient temperature changes, and a feedback arrangement ensures a linear meter reading. Such a measuring technique is referred to as 'true r.m.s.'.

Peak reading meters are sometimes required for transient or pulse height measurement. In theory a peak reading device can be used just as easily as an averaging meter for pseudo r.m.s. measurement of a sine wave, provided the scale factor is adjusted.

This is not good practice, however, because quite short interfering transients occurring near the peak can cause a large increase in the reading even though the r.m.s. and average values are virtually unaffected.

circuit design

Although there exist several possible circuit arrangements which are satisfactory for meter driving, the feedback method has much to recommend it for both d.c. and a.c. applications.

circuit for d.c. meter drive

Figure 15.1 shows two forms, Fig. 15.1a having a low but well-defined input resistance and Fig. 15.1b a high but ill-defined input resistance.

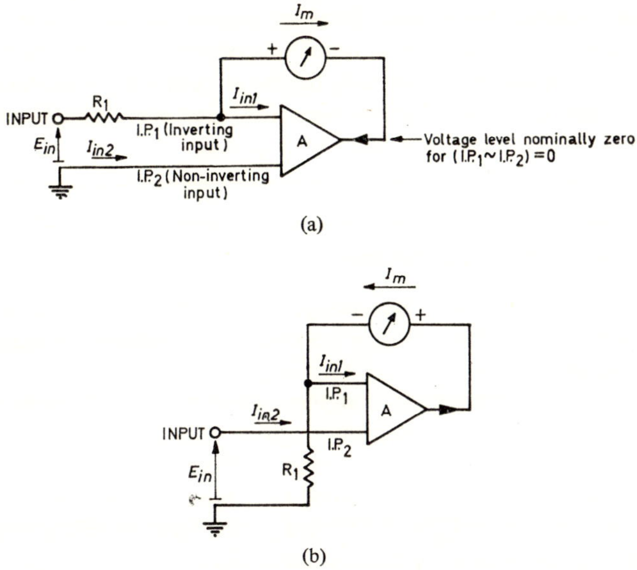

Fig. 15.1 (*a*) Meter drive circuit (low input resistance); (*b*) meter drive circuit (high input resistance)

The principle, already described in Chapter 13, is that throughout the linear operation of the amplifier in either circuit, the voltage on I.P.$_1$ will be almost equal to that on I.P.$_2$. If this were not the case, then the amplifier would drive to the limit of its capability in the direction tending to make the inputs equal, as in all ideal negative feedback amplifiers.

In Fig. 15.1a, I.P.$_1$ is therefore close to 0 V and a d.c. input E_{in} causes a current E_{in}/R_1 to flow. Provided the input resistance to I.P.$_1$ is high, this current flows directly into the meter, so that $I_m = E_{in}/R_1$.

In Fig. 15.1b, I.P.$_1$ is again at the same potential as I.P.$_2$, which in this circuit is E_{in}. A current E_{in}/R_1 therefore flows through R$_1$ to earth; as before, this current must flow through the meter if the input resistance to I.P.$_1$ is high, so that again $I_m = E_{in}/R_1$.

In either arrangement it can be seen that within the approximations made meter reading is unaffected by changes in meter resistance. The effective source resistance to the meter is very high even though the resistance between either meter terminal and ground is quite moderate.

The high-input-resistance form of circuit shown in Fig. 15.1b is generally preferred as it is usually inconvenient for the source to supply all the current required by the meter as in Fig. 15.1a. An input resistor can easily be added between I.P.$_2$ and earth should a lower but well-defined input resistance be required.

design points for d.c. meter drive

Although the basic arrangement given in Fig. 15.1b is simple, the design of the amplifier still requires care if the assumptions made are to be valid.

The first requirement is that the output current available from the amplifier must be sufficient to drive the meter in the chosen direction, or both directions if a centre-zero meter is to be used.

The input current I_{in1} required by I.P.$_1$ must be much less than the meter current; since $I_m = E_1/R_1 \pm I_{in1}$, the error from this cause is directly the percentage of I_{in1} relative to I_m.

The input current I_{in2} required by I.P.$_2$ is much less important; in Fig. 15.1a it has no effect at all, and in Fig. 15.1b it represents a load on E_{in}.

The gain required is calculated as for any negative feedback amplifier, i.e. the open loop voltage gain must exceed the over-all closed loop gain by the reciprocal of the allowable error. Almost any normal moving coil meter has a voltage drop in the region of 100 mV at f.s.d. so that at an early stage of design the order of gain required can be estimated. For example, if f.s.d. of 1 mA is to be obtained for 500 mV input, R_1 should be $E_{in}/I_m = 500\ \Omega$. Meter resistance will be about $100/I_m = 100\ \Omega$, so for a 1 per cent error (assumed to be caused only by insufficient gain) the gain of the amplifier itself should be $100(R_1 + R_m)/R_1 = 120$, *when measured with the meter load present.*

The amplifier must be designed so that the nominal standing level of the output is zero if the two inputs are at the same potential. If this condition is not obeyed then with zero input there will be a reading on the meter.

In a practical amplifier such a reading will inevitably exist and it is the designer's task to ensure that the resulting error is within the specification.

The easiest way to calculate the consequences of an imperfect amplifier is to assume zero output voltage from the amplifier and zero input to the complete circuit. Knowing the offset voltage and input current of the amplifier then ables the resulting meter reading to be deduced. In Fig. 15.1a, if the amplifier offset voltage for zero output voltage is $V_{o.s.}$, the current flowing in R_1 is $V_{o.s.}/R_1$ and this must flow into the meter giving this current reading. Naturally the output voltage of the amplifier must differ slightly from zero to deliver this current to the meter, but if A is high this is a negligible effect. Similarly any input current I_{in1} in Fig. 15.1a flows almost entirely through the meter since a path through R_1 would cause an input voltage change at I.P.$_1$ resulting in a very large change of E_{out}.

In Fig. 15.1b an input offset $V_{o.s.}$ again causes a current $V_{o.s.}/R_1$ to flow in R_1 and this current flows through the meter; an input current I_{in1} results in a current flow in the meter of I_{in1} such that none of this current flows in R_1.

practical d.c. meter-drive design

It is quite possible to use a linear microcircuit such as the M101 or μA 709 as the amplifier in these circuits. Often, however, the gain requirements for meter driving are modest and then a circuit of discrete components can be less costly. In particular one of the standard three-transistor amplifiers recommended in Chapter 13 is often suitable and this will be used in the following example.

Supposing it is required to drive a 1 mA, 100 Ω ± 10 per cent meter from a 1 V d.c. signal, the input resistance of the circuit to be more than 50 kΩ, and its maximum error to be ±2 per cent in addition to errors in the meter movement calibration.

The requirements for high input impedance dictates the circuit of Fig. 15.1b since the circuit of Fig. 15.1a would turn out to have an input resistance of 1 kΩ, which is the value for R_1 in either circuit ($R_1 = E_{in}/I_m = 1$ V/mA).

The gain required from the amplifier is of the order of 50 minimum when loaded by the total value of $R_1 + R_m$. The circuit described by Fig. 13.2 is therefore worth considering and a tentative design can be worked out.

Figure 15.2 shows the proposed meter drive, arranged to operate correctly on positive-going inputs only. The main design point is

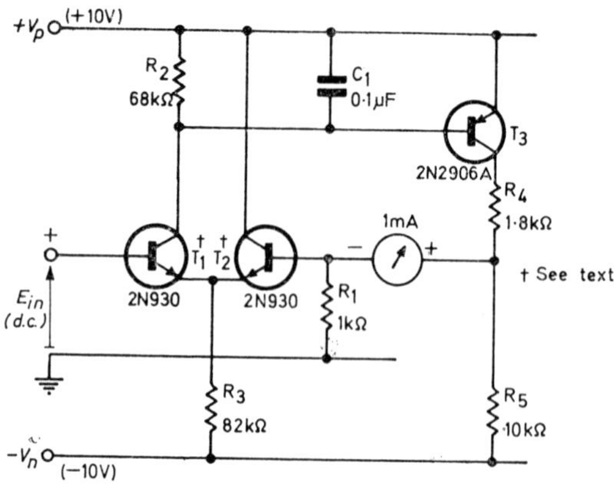

Fig. 15.2 Practical example (d.c.)

that T_3 must be capable of driving 1 mA into the meter and to achieve this T_1 and T_2 must operate at sufficient current. This being so (if $\beta_{min3} = 40$) the design is virtually complete, though some interesting details are worth mentioning.

Firstly, C_1 is so large that no RF oscillation can occur, yet so small that the meter response time is unaffected. Should it be desirable to slow the meter response then increasing C_1 is not recommended since it would result in different time constants according to whether the input became larger or smaller. This occurs because when T_1 turns on hard C_1 charges at a rate determined by the value of T_1 collector current ($dV/dt = i_{c1}/C$); when T_1 turns off, C_1 recovers exponentially through R_2 at a slower rate. A better method is to place a capacitor across the meter and, if necessary, to add first a resistor of several hundred ohms in series with the movement to increase the resulting time constant. If the movement, on the other hand, is itself over-sluggish, then a capacitor across R_1 has the effect of reducing the damping because the meter is overdriven for a time until the capacitor charges to the new value and allows the feedback loop to operate.

Secondly, the value of R_5 is non-critical if only one-way meter deflection is required. If a 1–0–1 mA, i.e. centre zero, meter is to be driven then R_5 has to supply 1 mA when E_{in} falls to -1 V. At that time T_3 collector is at about $-1{\cdot}1$ V and the limiting value for R_5 is $(10 - 1{\cdot}1)/10^{-3}$, i.e. 8·9 kΩ. In a real design R_5 would be perhaps 6·8 kΩ to avoid the need for tolerance calculations for V_n, R_1, R_m,

I f.s.d., R_5 and I_{CBO3}; and to ensure that on negative f.s.d. T_3 collector current is not allowed to fall below a few hundred microamps. Similarly, R_5 in the present design cannot be omitted since near zero input T_3 would be close to cut-off thus causing loop gain to fall. The actual value of R_5 could however range from, say 22 to 4·7 kΩ with little effect on circuit performance.

Thirdly, R_4 is essential (though often forgotten) to limit the meter current; without this precaution the meter could receive approximately ($\beta_3 \times \frac{1}{4}$) mA, e.g. 25 mA, if E_{in} reached +10 V. There are many circuits where it is impossible to protect directly against such excessive inputs without degrading performance: other means then have to be found, such as warning notices, input clamping circuits, or mechanical protection on the meter. In this case the protection by R_4 is simple, degrades performance by a negligible amount and should always be included. Bearing in mind paragraph 7 above, R_4 is designed to limit the meter current to 3 mA in the event of T_3 bottoming, and on the other hand, to be capable always of supplying 1 mA with no bottoming of T_3.

Fourthly, to meet the accuracy specification, the V_{be} of T_1 and T_2 have to be matched to better than ±10 mV if ±1 per cent error is permitted from this cause. Where this is to be held over a wide temperature range a dual transistor, e.g. 2N3680, may be required. For many applications it is permissible to use a 'zero set' potentiometer to set the meter reading to zero from time to time. Once the zero is set in this way an unmatched transistor pair (2N930, 2N2484, BC 108) can be expected to track within about 0·2 mV/degC so that a temperature change of 50 degC would give a 1 per cent error in this particular circuit. Figure 15.3 shows possible zero setting arrangements. The first (Fig. 15.3a) has the disadvantage that with V_{Re} large enough to give sufficient zero-setting, the circuit gain is noticeably reduced. The second has the snag that supply line variations affect the zero setting; also, R_7 plus a fraction of R_6 appears in parallel with R_1 in determining I_m/E_{in}, so that the over-all gain varies with the setting of R_6.

In spite of these difficulties both methods can be successful with careful design. For the present specification the method given in Fig. 15.3a is adequate (V_{Re} is chosen so that with all the tail current V_n/R_3 flowing in it the biggest discrepancy in V_{be} of T_1 and T_2 is exceeded, e.g. (10/82 kΩ × (V_{Re}) > 0·2 V, or V_{Re} > 1·64 kΩ). If the specification had required more gain then the method of Fig.

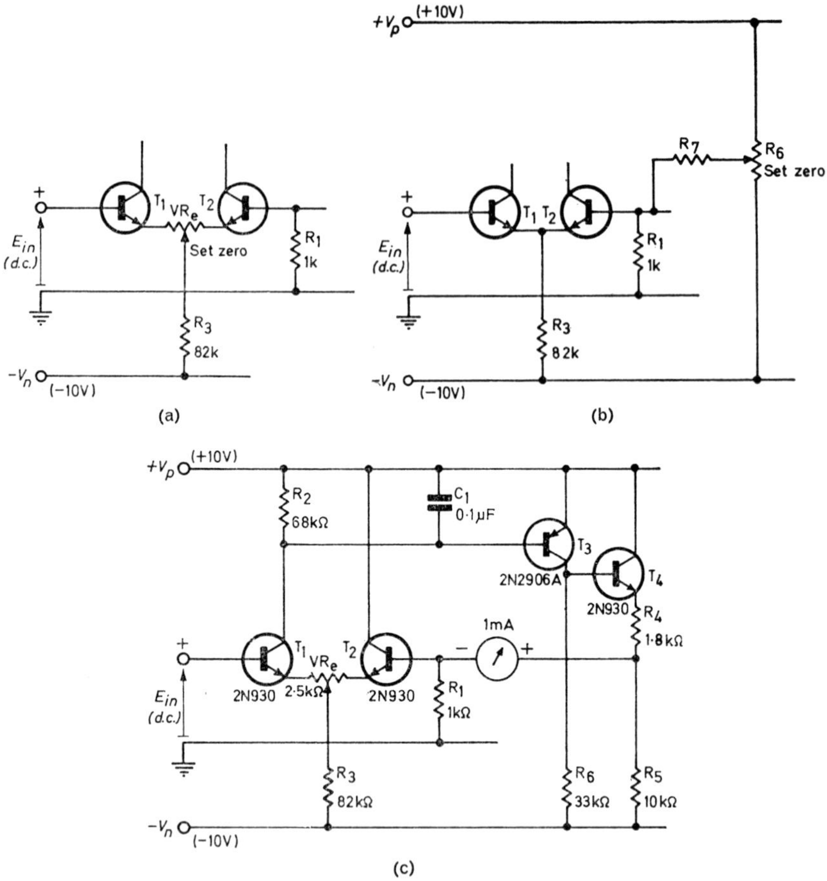

Fig. 15.3 (*a*) Zero set with emitter potentiometer; (*b*) zero set by base potentiometer; (*c*) use of emitter follower to restore lost gain

15.3b could be used, returning R_6 to Zener stabilised lines and ensuring $R_6 \ll R_7$. Alternatively, the circuit of Fig. 15.3c could be used, where the loss of gain caused by the emitter circuit potentiometer is made up by an output emitter follower.

Lastly, the input impedance requirement (greater than 50 kΩ) is very easily met since, even without feedback, the input impedance is already of the right order. With feedback present the value achieved is about 1 MΩ.

This again, as in Chapter 13, brings up the question of source resistance. When specifying accuracy this should be quoted; in particular when specifying the maximum permissible output for zero input it must be made clear whether the source is to be open circuit for this measurement. If this is so (a very common situation) then a much worse zero error will occur than if the source e.m.f. were merely reduced to zero. In Fig. 15.3a the zero error with a 'reasonable' source of 2·5 kΩ – consistent in that it allows a 50 kΩ input impedance to give a $\frac{1}{2}$ per cent error – is about $2{\cdot}5 I_{b1} \times 10^3$ equivalent input voltage, or 0·25 per cent of f.s.d., since $I_{b1} \approx 1\ \mu$A. On open circuit the zero error is difficult to calculate; T_1 has zero base current and is close to cut off. The resulting output error can be reduced by adding a resistor of 68 kΩ in parallel with the input. The input impedance is still more than 50 kΩ and the output for open circuit input differs from the zero source e.m.f. figure by at most $(I_{b1} \times 68 \times 10^3)$ 100 per cent of f.s.d., i.e. about 6·8 per cent. If this is unacceptable, the operating current of T_1 and T_2 must be reduced to, e.g. 10 μA and a circuit like Fig. 15.3c used to ensure sufficient meter drive.

Even severe variations in supply voltage have little effect on performance. The circuit fails at very low values of V_p when T_3 bottoms at maximum input and is still unable to drive the meter to f.s.d. When V_p is increased the protection given by R_4 is reduced so that overloads greater than 3 × f.s.d. can be delivered to the meter; at very high values of V_p, T_3 or T_1 will break down due to excessive V_{cb}. Typical V_p variations for which only slight loss of performance will result is +3·5 to +40 V. If V_n is reduced to a very low value, T_1 may be unable to drive R_2 and T_3 because of the reduction in tail current through R_3. When V_n is very large T_3 may be unable to drive the current now demanded by R_5 although it is assisted in this by extra drive from T_1 which now has extra tail current. Ultimately T_3 bottoms because of the drop across R_4. Typical V_n variation for slight performance change is 2 to 40 V, if V_p = 10 V. These enormous supply changes without great loss of performance also imply considerable immunity to ripple present on the supplies.

circuit for a.c. meter drive

The same basic arrangement is suitable for a.c. drive but then the design calculations usually become much simpler. Figure 15.4 illustrates the two feedback connections which correspond to the

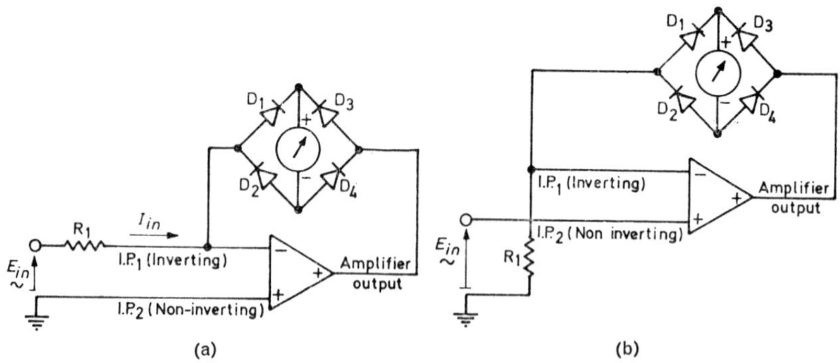

Fig. 15.4 (*a*) Basic a.c. meter drive – low input impedance; (*b*) basic a.c. meter drive – high input impedance

d.c. circuits of Figs 15.1a and 15.1b. The first configuration operates with a sine wave input as follows. $I.P._1$ remains near zero potential for all input signals within the capability of the circuit, so that the input impedance is approximately R_1 and $I_{in} \approx E_{in}/R_1$; this current must flow to the meter bridge provided the a.c. input current required by the amplifier is negligible. On positive input half cycles the current can flow only through D_1, then through the meter in a downwards direction, and then through D_4, to the amplifier output. On negative half cycles the current flows from the amplifier output through D_3, through the meter (again downwards) and through D_2 to R_1.

The meter therefore receives a full-wave-rectified current waveform of average value $(2/\pi)(\hat{E}_{in}/R_1)$, the current being independent of the meter resistance and of the forward voltages dropped by the diodes, provided the loop gain is adequate. A similar situation exists in Fig. 15.4b but here the input resistance is high as in its d.c. counterpart Fig. 15.1b.

design points for a.c. meter drive

As shown the circuits of Fig. 15.4 are also sensitive to d.c. inputs which would deflect the meter in the same direction regardless of input polarity. For certain applications this combination which transforms $(|E_{in\,d.c.}| + E_{in\,a.c.})$ into a meter current could conceivably be useful but in general separate a.c. and d.c. measurements are made.

The circuits of Fig. 15.4 could be made insensitive to d.c. inputs merely by capacitively coupling E_{in} and adding a resistor to ground from I.P.2 in Fig. 15.4b. However, a meter reading would then exist for zero input owing to amplifier offsets. This may be acceptable, depending on the properties of the amplifier. If not, then simply blocking the direct current from the meter circuit would be insufficient because this would break the d.c. negative feedback loop. This would result in the amplifier limiting if the product of its gain

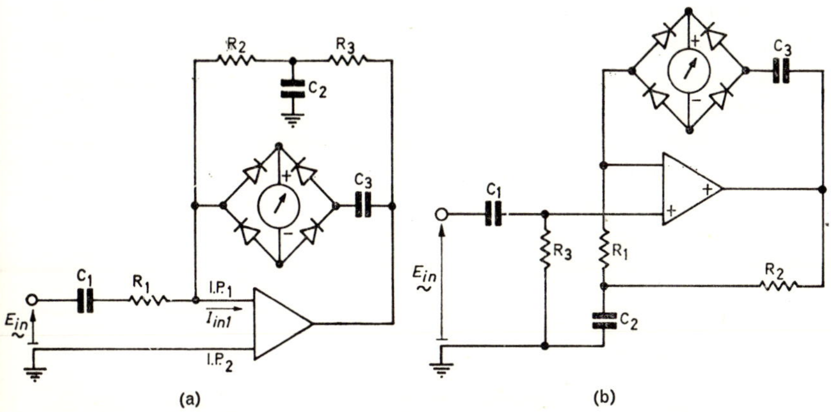

Fig. 15.5 Removing d.c. from meter readings: (*a*) low input impedance circuit; (*b*) high input impedance circuit

and input offset should exceed its possible output swing—almost a certainty in a practical amplifier of sufficient gain for this application. A logical solution is given in Fig. 15.5a, where the d.c. loop is completed by R_2 and R_3 and is decoupled at signal frequencies by C_2. The value of C_2 is not critical but too low a value means that the meter is driven by a lower impedance and that a noticeable proportion of I_{in} flows into R_2 because the signal across R_2 is no longer close to zero. The choice of C_3 is much less important and to a first order has no influence on meter reading. The ultimate result of too low a value here is that the required voltage swing at the amplifier output becomes large for a given meter current and this can cause limiting. C_1 on the other hand directly affects circuit gain and must have low reactance compared with R_1 at the lowest signal frequency. R_2 and R_3 are again not critical; too low a value for R_3 would require excessive C_2 for good decoupling and would represent a heavy load on the amplifier; too high a value for $(R_2 + R_3)$ could result in an output standing level (which is equal to $V_{o.s.} + I_{in1}(R_2 + R_3)$) too near the limiting level for the amplifier.

The alternative and more useful circuit of Fig. 15.5b uses the same principle of separate d.c. and a.c. feedback loops. Here C_2 is more critical than in Fig. 15.5a since it performs the function of decoupling the gain determining resistor R_1; it must therefore be calculated in the same way as C_1 in Fig. 15.5a. R_2 represents an excessive load on the amplifier if too low in value, or too large an output offset ($\pm V_{o.s.} I_{in1}(R_1 + R_2)$) if too high. R_3 is similarly calculated – it adds $I_{in2}R_3$ to the output offset and also determines the input impedance. C_1 must couple adequately at the lowest signal frequency.

In either circuit, the d.c. offset of the amplifier has no first-order effect on performance provided no amplifier limiting takes place when the a.c. signal is applied. This means that C_1 may be omitted if the d.c. component of E_{in} is small.

choice of diodes

An alternative form of meter connection often used is to replace D_1 and D_2 (Fig. 15.4) by capacitors. The saving from the point of view of price versus bulk is doubtful when large capacitors have to be

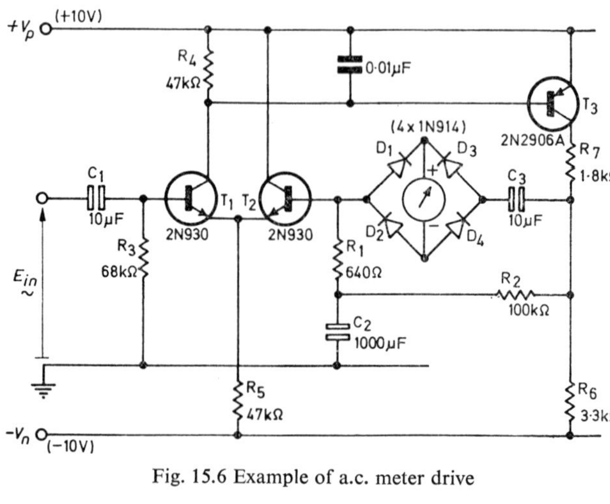

Fig. 15.6 Example of a.c. meter drive

used for low-frequency operation. Meter response is slowed down as it is shunted by the capacitors, and circuit gain is halved, i.e.

$$I_m = \frac{1}{\pi} \cdot \frac{\hat{E}_{in}}{R_1}$$

The diodes must, in all circuits considered here, have reverse leakage which is negligible compared with meter f.s.d. current; this usually dictates the use of silicon devices. The large forward drop which these diodes possess is fortunately not important owing to the feedback mechanism.

practical a.c. meter drive design

As in the d.c. drive circuits, a linear microcircuit is often ideally suited for use in the a.c. circuits shown, especially when a sensitive measurement requiring high amplifier gain is to be made. For comparison with the d.c. example, however, the simple three-transistor amplifier will be used again and a similar specification used.

In this example the source is to be a 1 V peak, sinusoidal waveform in the band 50 Hz to 1 kHz. Input resistance is to exceed 50 kΩ; the maximum error to be ±2 per cent, and the meter, as before is 1 mA f.s.d., resistance 100 Ω ± 10 per cent.

Figure 15.6 gives a suitable circuit, using the three-transistor amplifier with the circuit of Fig. 15.5b. As before the input impedance at T_1 base will be many times the required value and R_3 may be 68 kΩ, giving $1/\omega C_1 \ll 68 \times 10^3$ when $\omega = 2\pi \times 50$, e.g. $C_1 = 10$ μF. R_1 is calculated from I_{av}, i.e. meter reading $= (2/\pi)\hat{E}_{in}/R_1$, giving $R_1 = 640\ \Omega$. C_2 is now given by $1/\omega C_2 \ll 640$ when $\omega = 2\pi \times 50$, e.g. $C_2 = 1{,}000$ μF. $(R_2 + R_1)$ must be such that $I_{b2}(R_2 + R_1)$ gives an output offset well clear of limiting, e.g. less than 1 V, giving $R_2 < 470$ kΩ, e.g. 100 kΩ (there is nothing to be gained by a higher value since at 100 kΩ its load on the output is negligible). C_3 should drop, say less than 1 V pk at f.s.d., i.e. $1/(R_1 \omega C_3) \leqslant 1$ when $\omega = 2\pi \times 50$, giving $C_3 = 10$ μF.

Note that R_6 must be capable of driving at least $\pi/2$ mA on negative input half cycles as T_3 tends towards cut off and that at this time V_{C_3} has fallen more than in the corresponding d.c. circuit (Fig. 15.2). This is due to the forward drops of the two diodes D_1 and D_4, plus (algebraically) the drop across C_3, and to the factor relating mean to peak. When E_{in} reaches -1 V at its negative peak, T_2 base also falls to -1 V, the $C_3/D_3/D_4$ junction falls to -1 V + 2 diode drops + meter drop, i.e. about $-2{\cdot}5$ volts, and T_1 collector falls to $-(2{\cdot}5^2 + V^2{}_{c3})^{\frac{1}{2}}$, i.e. $-2{\cdot}6$ V. The limit value for R_6 is therefore $(2/\pi)(10 - 2{\cdot}6)/10^{-3}$, or 4·7 kΩ and in order to allow generous margins for diode tolerance, -10 V supply tolerance, etc. and to ensure that T_3 is not near cut off, a value of 3·3 kΩ is reasonable.

On positive excursions it is easily seen that this resistor will pass (12·6/3·3) mA, i.e. 3·8 mA, so that T_3 is then providing in all $(3{\cdot}8 + \frac{1}{2}\pi)$ mA, i.e. 5·4 mA. Its base drive should therefore exceed (5·4/40) mA, i.e. 135 μA, so the emitter current for T_1 and T_2 has been made nominally 200 μA, giving $R_5 = 47$ kΩ.

The generous margins allowed on d.c. conditions avoid the need for any accurate calculations and have in this case resulted in suitable circuit values. If the above had resulted in T_1 and T_2 running at say 6 mA (as would have occurred with a 30 mA meter) then the output offset due to $I_{b2}(R_1 + R_2)$ and $I_{b1}R_3$ would be excessive. An attempt would then be made to calculate the margins more finely but, inevitably, an extra transistor would have to be introduced as in Fig. 15.3c.

R_7 is again used for meter protection and is designed to limit the peak current to 4·5 mA, giving a value of greater or equal to (10 − 2·6)/4·5 kΩ, or 1·8 kΩ. R_4 is determined as usual by I_{CBO} considerations and has a safe value of 47 kΩ. C_4 is large enough to represent a dominant time constant in the feedback loop but does not appreciably affect 1 kHz performance.

practical hints

Apart from the tests to which any amplifier would be subjected it is good practice to check the meter driving impedance by adding a resistor equal to the meter resistance directly in series with the movement. The drive impedance is readily calculated from the change in meter reading.

In a.c. meter drives the waveforms at the amplifier output and at critical signal points such as T_2 base in Fig. 15.6 should be examined. In that circuit T_2 base voltage should be a replica of E_{in} and the output voltage at the $C_3/D_3/D_4$ junction should be square with a superimposed sine wave (Fig. 15.7a). This is caused by the change-

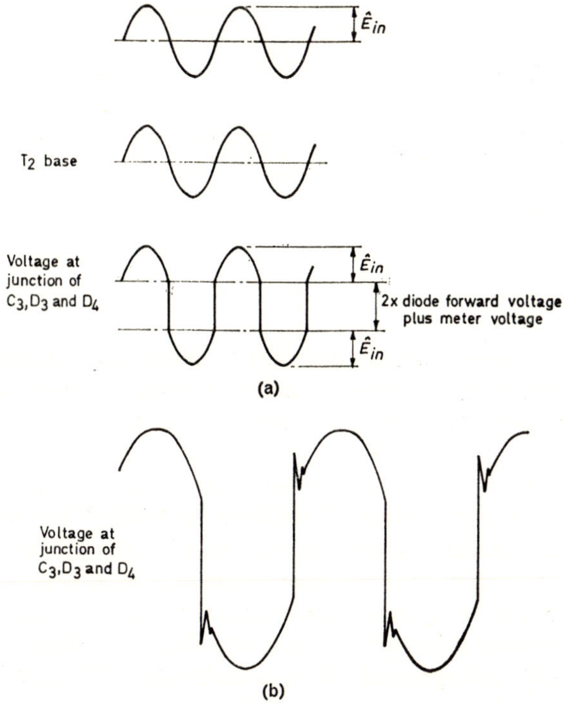

Fig. 15.7 (*a*) Waveforms for circuit of Fig. 15.6; (*b*) effect of meter inductance

over of conduction of the two parts of the diode bridge and the shock excitation thus provided often provokes a ring due to meter inductance Fig. 15.7b. This ring can reach alarming proportions if the meter connections are long and can result in the amplifier limiting in one direction thus changing the reading. It is always advisable, therefore, to add a capacitor, e.g. 0·1 μF disc ceramic, in parallel with the meter at the diode-bridge end of long meter leads. This reduces the ring to a negligible level and thus prevents reading errors caused by limiting and radiation from the meter lead to other circuits.

CHAPTER 16

precise constant current sources

CONSTANT current sources are used to define accurate currents in ramp generators, to serve as collector loads in ultra-high-gain amplifiers (see Chapter 12 of EDH), as long tails in differential amplifiers and to provide constant operating currents in the presence of large signal or supply voltage variations. In most applications it is essential that one terminal of the load be 'earthy', i.e. connected directly to a power line or to earth, and this rules out certain configurations using over-all negative feedback. The driving of moving coil meters and direct acting pen recorders requires a special form of high impedance drive for both a.c. and d.c. input signals; this requires a different approach and was dealt with in Chapter 15.

standard circuit

Figures 16.1 and 16.2 show the usual forms of constant current circuit. In each case the load current is largely independent of the value of R_L, i.e. the circuit has high output impedance. In Fig. 16.1 the actual value of load current is determined mainly by V_n, R_1, R_2

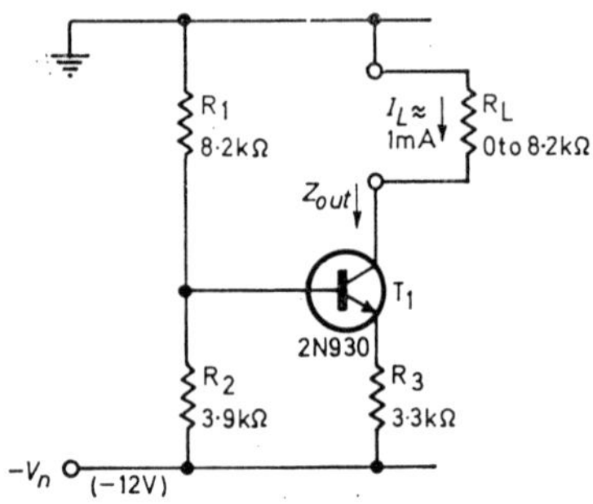

Fig. 16.1 Constant current circuit

and R_3, and in Fig. 16.2 the current depends principally on V_z and R_3 and is thus virtually independent of V_n and R_1.

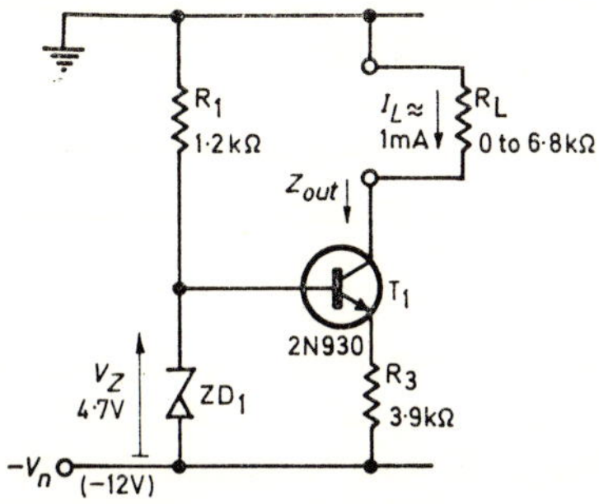

Fig. 16.2 Stabilized constant current circuit

It is easy to show (see Appendix 3, p. 270 of EDH) that the output impedance Z_{out} of the above circuits varies from r_c/β to r_c (using T parameters) as R_3 varies from zero to infinity (there being little further change in Z_{out} when R_3 exceeds a few kilohms).

imperfections of the standard circuits

When the circuit of Fig. 16.1 is used as an emitter circuit 'long tail' (Fig. 16.3) the actual value of the total emitter current I_1 is not so important as the magnitude of Z_1 and for good rejection of in-phase

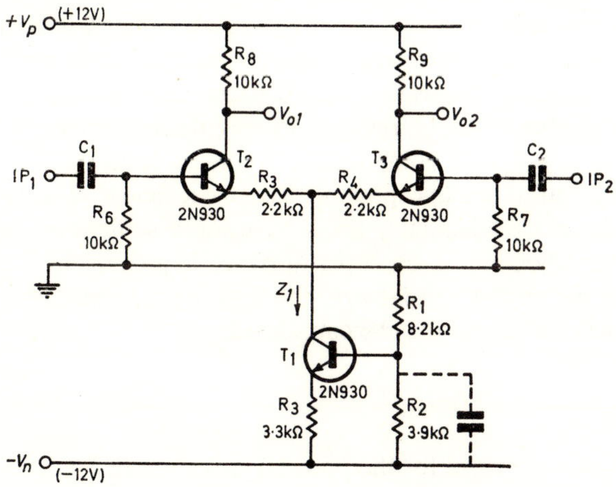

Fig. 16.3 Differential amplifier with 'long tail'

signals applied to I.P.1 and I.P.2, $Z_1 \gg 2{\cdot}2\ \mathrm{k}\Omega$. This subject is dealt with in more detail on p. 86 of EDH.

It is therefore unnecessary to use the Zener diode shown in Fig. 16.2 from the point of view of current stability, although the danger now exists of modulation of the defined current by ripple on the supply line. Where this is important, either a Zener diode should be used or R_2 should be decoupled by a capacitor to the $-V_n$ line, i.e. across R_2, not to earth. At high frequencies the collector base capacitance of T_1 lowers the magnitude of Z_1. This begins to reduce circuit performance at surprisingly low frequencies; with $C_{cb} = 10$ μF, the effective value of Z_1 at 100 kHz is only 150 kΩ, whilst the low-frequency value of Z_1 may be several megohms.

In a video and sometimes even an audio amplifier this implies a push-push signal rejection which deteriorates rapidly as the signal traverses the frequency band.

Different problems arise when an accurate current value is to be defined, as in the generation of a precise ramp function. The circuit of Fig. 16.2 is then normally used since it reduces current variations caused by changes of supply voltage and by base current variations in T_1, since the base circuit impedance is low.

Although this gives adequate performance in many applications, the defined current varies with temperature from five causes: Zener diode voltage variation (caused by R_1 variations, variation in I_b and the Zener temperature coefficient); R_3 temperature coefficient; V_{be} temperature coefficient; transistor I_{CBO} transistor α variation (so that I_L is not a fixed percentage of I_e); α rather than β is used in this chapter since the performance depends more directly on collector current/emitter current rather than collector current/base current.

Some of these may be reduced by any desired extent by simply specifying appropriate components. The main contributors then are α and V_{be} variations and possibly Zener coefficient. The latter can usually be made adequately small by using a sufficiently expensive diode, but when only a small performance improvement is required to make the circuit satisfactory, the approximate matching of V_{be} and Zener coefficients may be considered.

remedies

The only simple cures for the poor HF rejection in Fig. 16.3 are the choice of, above all, a low C_{cb}, high f_T transistor type for T_1 and to operate at the highest reasonable V_{cb} voltage (which gives a low value for C_{cb}). If there is a high voltage negative line available the best cure of all is to replace T_1, R_1, R_2 and R_3 by a resistor, whose HF performance is likely to be much superior to the circuit under discussion!

Much more can be done to improve the accuracy of Fig. 16.2 at zero frequency. As proposed earlier, the matching of Zener and V_{be} coefficients improves temperature performance. This is hardly practical for really accurate work but the principle is good and one should, as a habit, favour a low voltage Zener in this circuit, e.g. 4·7 or 5·1 V. The temperature coefficient is then negative and tends to track V_{be} roughly. (Conversely, in the stabilizer circuit shown in Fig. 1.8, or 6.2 or 6.8 V Zener is appropriate because in the emitter circuit a positive coefficient is required for this purpose.) In common with many other error cancellation techniques this rough and ready method of compensation may just tip the scales favourably when a design is only marginal.

Variations in α become significant when the output current is to remain with a few per cent of the ideal value; I_{CBO} is significant only when I_L is very small, e.g. below 100 μA, and the operating temperature is high (>100°C), assuming a silicon transistor is used.

When Zener diode, V_{be} and coefficients combine to make the design out of tolerance by an order of magnitude, the circuit of Fig. 16.4 should be considered. It is identical in most respects to a voltage stabilizer circuit (see, for example, p. 208 of EDH), but the

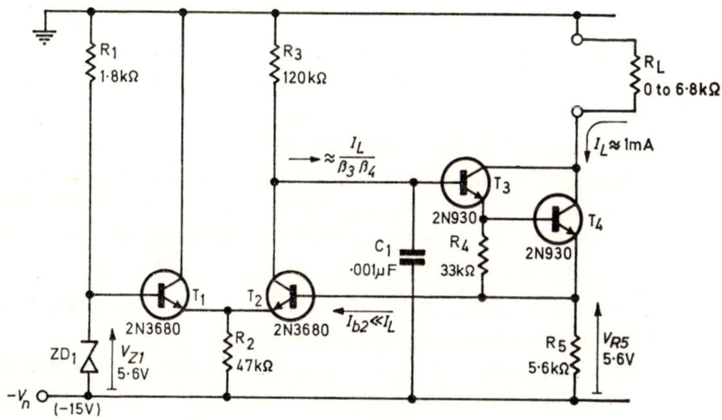

Fig. 16.4 Four-transistor constant current circuit

output is taken between the T_3, T_4 collector junction and the earth return. The idea is that the voltage across R_5 will be equal to V_{Z_1} so that the current in R_5 is V_{Z_1}/R_5. Because of the manner of connection between T_3 and T_4 (note that T_3 collector is not connected to the zero line but to T_4 collector, and that R_4 is returned to T_4 emitter and not to $-V_n$), almost all of this current passes into the load, only about $1/\beta_3\beta_4$ of it being lost.

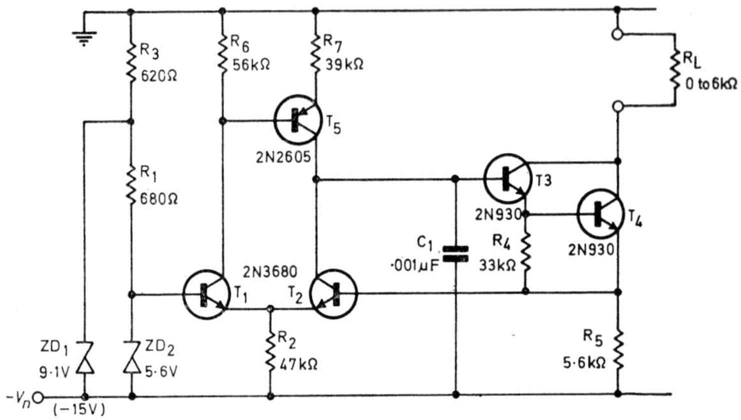

Fig. 16.5 Development of circuit of Fig. 16.4

Although the output current is still directly dependent upon Zener voltage V_{Z_1}, most of the other effects are much reduced. The V_{be} of T_1 and T_2 can be bought matched to any desired extent in initial value and temperature coefficient, and R_5 can be as accurate as the particular application demands. The V_{be} of T_3 and T_4 are insignificant to a first approximation, being within the feedback loop. They can be made even less significant by using a constant current collector load for T_2 (see Chapter 12 of EDH), and Zener performance can be improved by pre-regulation (Fig. 16.5). Clearly, the circuit is capable of further development in the form of extra loop gain, e.g. by the use of a field effect device instead of T_3 and T_4 (see Fig. 16.6; note ZD_3, or R_8 is essential to allow adequate gate bias for F.E.T.1 without causing T_2 to bottom).

In comparison with Fig. 16.2, the circuit of Fig. 16.4 therefore enables V_{be} to be balanced out very accurately and makes the effect of β variations negligible.

design of improved constant current circuit

For Fig. 16.4, design begins with the choice of ZD_1, usually near 5·6, 9 or 11 V, depending on how close the tolerance has been achieved by the manufacturer (see p. 7 of EDH). R_1 is designed for

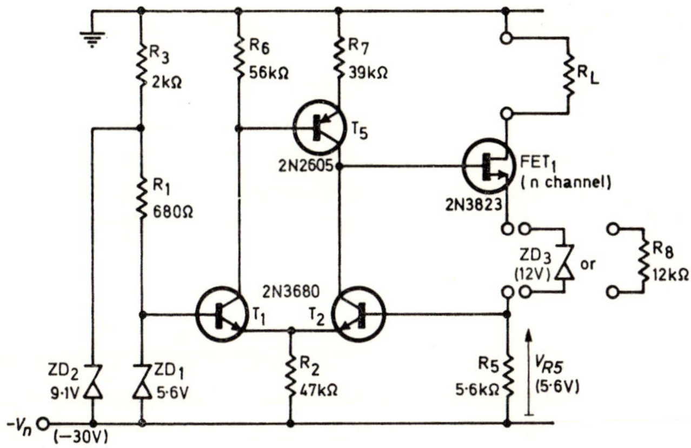

Fig. 16.6 Use of F.E.T. instead of T_3T_4

correct nominal current for ZD_1 and R_5 is then calculated to give correct load current $I_L = V_{Z_1}/R_5$ and a suitable type of resistor chosen. R_4 is the 'I_{CBO} resistor' (Chapter 1) and is usually 10 to 100k Ω. Since T_3 base current will be small ($<10\ \mu$A) in most applications the current in T_1 and T_2, governed by R_3, can normally be chosen to be optimum for T_1 and T_2 V_{be} matching—usually between 10 and 500 μA. T_1 and T_2 may be separate or dual according to the degree of matching required, e.g. 2N3680 dual, and T_3 and T_4 should be high β types with adequate voltage and current rating.

The practical limit to the accuracy obtainable is usually set by component cost; by using for example the BZX47 (Mullard) Zener diode with a coefficient of $\pm 0{\cdot}0005\%$/degC, a 2N3680 (Texas) dual transistor for T_1 and T_2 and a metal film resistor for R_1 and R_6, the temperature coefficient for I_L can be as small as 0·002%/degC, although part of R_5 will have to be variable to accommodate initial tolerance on V_{Z_1}. If still better temperature performance is needed, the critical components can be mounted in a small oven of the type intended for crystal control. An inexpensive oven will stabilize the temperature to ± 2 degC for outside temperatures down to 75 degC below the thermostat setting (typically 0 to 75 degC). A high grade oven costing tens of pounds will stabilize to a fraction of a degree.

adjusting the load current

It is often required to adjust the output current by means of a variable resistor or in proportion to an applied voltage. In Fig. 16.1 this can be achieved by varying R_3 or R_1/R_2; in Fig. 16.2 by varying R_3, and in Fig. 16.4 by varying R_5. When temperature drift is important the variation of R_3 in Fig. 16.2 would be unsatisfactory because as T_1 emitter current is changed, the temperature coefficient of its V_{be} also changes and no longer matches the Zener coefficient.

The improved circuit is therefore superior for variable as well as fixed currents; another satisfactory method of adjustment (Fig. 16.4) is to add a potentiometer across Z_{D_1} and connect T_1 base to the slider. If the output current is to be varied in sympathy with an applied voltage then T_1 base may be taken to this voltage and Z_{D_1} removed. When either of these methods is used R_3 may have to be returned to a more negative supply than $-V_n$ if current adjustment to low levels is required.

It is particularly to be noted that none of these circuits is satisfactory for output currents that approach zero, let alone those which are required to reverse direction. Much extra circuitry is required to achieve this using the same operating mechanism.

Fortunately, the most common use for such a circuit is in driving meters or direct pen recorders in which it is usually permissible to 'float' the meter coil so that neither terminal is earthed. This enables a simpler method to be used, which was fully described in Chapter 15.

bibliography

Although many of the following do not deal directly with transistor circuits, they help the designer by their approach to related design problems. Littauer's work is particularly well written.

Angelo, E. J. (1965). *Electronic Circuits.* London and Philadelphia; McGraw-Hill.

Bode, H. W. (1945). *Network Analysis and Feedback Amplifier Design.* London; Van Nostrand.

Cain, W. D. (1969). *Engineering Product Design.* London; Business Books.

Cattermole, K. W. (1965). *Transistor Circuits.* London; Heywood.

Dammers, D. G., Haantje, J., Otte, J. and van Suchtelen, H. (1950). *Electronic Valves,* Vol. 4. London; Philips Technical Library.

Deketh, J. (1950). *Electronic Valves,* Vol. 1. London; Philips Technical Library.

Dunster, D. F. (1969). *Semiconductors for Engineers.* London; Business Books.

Hardie, A. M. (1964). *Elements of Feedback and Control.* London; Oxford University Press.

Hemingway, T. K. (Second edition, 1970). *Electronic Designer's Handbook.* London; Business Books.

Heydroe, P. J. (1950). *Electronic Valves,* Vol. 7. London; Philips Technical Library.

Jones, D. D. and Hilbourne, R. A. (1957). *Transistor A.F. Amplifiers.* London; Iliffe.

Lewis, I. A. D. and Wells, F. H. (1959). *Millimicrosecond Pulse Techniques.* Oxford; Pergamon.

Littauer, R. (1965). *Pulse Electronics.* London and Philadelphia; McGraw-Hill.

Massachusetts Institute of Technology (1943). *Radiation Laboratory Series,* Vols. 1–28. London and Philadelphia; McGraw-Hill.

Middlebrook, R. D. (1963). *Differential Amplifiers.* New York; Wiley.

Nyquist, H. V. 'Regeneration Theory' from *Bell System Technical Journal,* Vol. 11, Jan. 1932, pp. 126–147.

Philips, Eindhoven (1950). *Electronic Valves,* Vol. 3. London; Philips Technical Libraries.

—— (1950). *Electronic Valves,* Vol. 3a. London; Philips Technical Libraries.

Philips, Eindhoven (1950). *Electronic Valves*, Vol. 5. London; Philips Technical Libraries.

—— (1950). *Electronic Valves*, Vol. 6. London; Philips Technical Libraries.

Reintjes, J. F. and Coate, G. T. (1952). *Principles of Radar*. London and Philadelphia; McGraw-Hill.

Sturley, K. R. (1965). *Radio Receiver Design*, Vol. 1. London; Chapman and Hall.

— (1965). *Radio Receiver Design*, Vol. 2. London; Chapman and Hall.

Wolfendale, E. (Editor) (1958). *The Junction Transistor and Its Application*. London; Heywood.

index